Fractals
the secret code of creation

Jason Lisle

First printing: June 2021

Copyright © 2021 by Jason Lisle. All rights reserved. No part of this book may be used or reproduced in any manner whatsoever without written permission of the publisher, except in the case of brief quotations in articles and reviews. For information write:
Master Books®, P.O. Box 726, Green Forest, AR 72638

Master Books® is a division of
New Leaf Publishing Group, Inc.

ISBN: 978-1-68344-240-0
ISBN: 978-1-61458-780-4 (digital)
Library of Congress Number: 2021938434

Designed by Left Coast Design

Unless otherwise noted, Scripture taken from the NEW AMERICAN STANDARD BIBLE®, Copyright © 1960, 1962, 1963, 1968, 1971, 1972, 1973, 1975, 1977, 1995 by The Lockman Foundation. Used by permission.

Please consider requesting that a copy of this volume be purchased by your local library system.

Printed in the United States of America

Please visit our website for other great titles:
www.masterbooks.com

For information regarding author interviews, please contact the publicity department at (870) 438-5288

MORE FASCINATING FRACTALS WITH DOWNLOADABLE SOFTWARE!

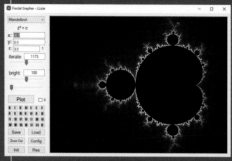

Dr. Jason Lisle takes the reader on a unique journey through fractals and how complex, beautiful, and amazing they are in *Fractals: The Secret Code of Creation*. Now you can explore fractals further and print your own images with Fractal Grapher©, an included program developed by Dr. Lisle of Biblical Science Institute (available for Windows®). It allows you to

- Explore fractals as you plot not only Mandelbrot sets, but nine other types of sets
- Save images as *.png, *.bmp, *.jpg, *.tiff, or *.gif
- Change brightness, contrast, and colors, and even import your own custom palette
- Continue where you left off by loading saved images back into the program
- Control plotting speed when zooming in to larger iterations and more complex details
- Learn tips on improving the resolution of your images
- Download a helpful set of instructions with screenshots that gives details and advice for you to successfully explore the extraordinary world of fractals.

To download your copy, visit:
https://biblicalscienceinstitute.com/FractalGrapher/

> The Images in this book are not the result of human creativity. The only human element is the selected color scheme, but the shapes stem from the mind of God.

Table of Contents

The Secret Code	4
Valley of the Seahorses	26
Valley of the Double Spirals	38
Infinite Elephants	52
Scepters on Seahorses	62
Changing the Formula	70
Multibrots	90
Fractals and the Christian Worldview	116
The Tricorn	136
The Sine Function	150
Geometric and 3D Fractals	170
Physical Fractals and the Grand Dilemma	194

Dr. Jason Lisle is a Christian astrophysicist who writes and speaks on various topics relating to science and the defense of the Christian faith. He graduated from Ohio Wesleyan University where he majored in physics and astronomy and minored in mathematics. He then earned a master's degree and a Ph.D. in astrophysics at the University of Colorado in Boulder. Dr. Lisle began working in full-time apologetics ministry, specializing in the defense of Genesis. His most well-known book, *The Ultimate Proof of Creation*, demonstrates that biblical creation is the only logical possibility for origins.

The Secret Code

Those who reject God like to explain the complexity of biological life by appealing to Darwinian evolution — the gradual changing of more primitive forms into more advanced forms as the unsuccessful cases are eliminated.

What if mathematicians discovered a secret code embedded in math itself? What would that mean? Suppose that when analyzing certain sets of numbers, we found an amazing work of art embedded in them, far more intricate and complex than any work of man. How would we make sense of such complex beauty in something as simple as numbers? Who put it there?

In fact, just such a code of astounding beauty was discovered in the 1980s. Artwork of tremendous beauty and infinite complexity had been hidden in numbers from the beginning of time. Yet it lay undiscovered until computer technology had advanced to the point that otherwise tedious computations could be performed with rapid efficiency. The beautiful images in this book are not the work of man. They are the very images that were discovered in sets of numbers, hidden in plain sight. How can we make sense of this? Who or what is responsible for these amazing shapes?

I suggest that secular thinking has no answer. Those who reject God like to explain the complexity of biological life by appealing to Darwinian evolution — the gradual changing of more primitive forms into more advanced forms as the unsuccessful cases are eliminated. This view has its problems, of course, but my point is that such an explanation is not even plausible for numbers because numbers do not evolve. It is not as though the number 7 gradually evolved from the number 3. Numbers have always been what they are. Therefore, the artwork displayed in this book did not evolve. It has always existed, being built into numbers.

I suggest that the Christian worldview alone can make sense of this secret code built into numbers. As such, the images in this book are a demonstration of the truth of the Christian worldview. The same God who built beauty into the physical world has also built beauty into the abstract world of numbers. We don't often think about God creating numbers. We tend to think of God creating physical things.

But even abstract conceptions like numbers could not exist apart from God.

Numbers are a concept of quantity. As concepts, they exist in the mind. We can represent numbers with a written numeral like the Arabic numeral "2" or the Roman numeral "II." But these are merely physical representations of an idea. After all, erasing the physical symbol "2" will not cause the number 2 to cease to exist! The number itself is abstract; it cannot be touched or seen, but it exists as a concept of the mind.

We can think about numbers in our mind, but we did not create them or the rules pertaining to them. It is not as though some ancient human simply decided to invent the number 2 and arbitrarily decreed that 2 + 2 should equal 4. No. Humans *discovered* numbers and the relationships between them. That means that numbers and the relationships between them existed before humans. This makes sense in the Christian worldview because numbers existed in the mind of God from the beginning of time. The mind of God is responsible for the existence of numbers and the rules governing their relationships. It has been the privilege of human beings to discover these rules by the gift of logical reasoning that the Lord has so graciously given. When we discover a mathematical truth, we have discovered something about the way God thinks.

The images in this book therefore represent an infinitesimal glimpse into the mind of God. God's thinking is not only flawlessly rational, but supremely beautiful as well. But exactly how were these images discovered? Where do they come from? A little background information will be helpful.

> When we discover a mathematical truth, we have discovered something about the way God thinks.

THE SECRET CODE 7

Sets

The images in the pages of this book are maps of sets of numbers. A set of numbers is just what you think it is: a group of numbers that have something in common. There are all kinds of sets. Most sets include some numbers and exclude others. Consider the set of even numbers. This set includes numbers like 2, 4, 6, 8, 10, but excludes numbers like 1, 3, 5, 7, 9. The set of negative numbers includes numbers like -3, -4, -5, -1/2, but excludes numbers like 2, 5, 7, π, and so on. You can even have the set of all numbers, which includes everything and excludes nothing. You can also have an empty set, which includes no numbers at all.

In sets like those mentioned above, you can tell if a number belongs or not just by looking at it. You know the number 24,389 does not belong in the set of even numbers because even numbers always end in 0, 2, 4, 6, or 8. You know that 57 does not belong in the set of negative numbers because there is no negative sign in front of it. But with some sets, you cannot tell just by looking at the number if it belongs or not. You have to do some work.

Consider the set of prime numbers: those natural numbers that cannot be formed by the product of two natural numbers other than themselves and 1. Does the number 14,351 belong to this set? You probably cannot tell just by looking at it. You have to do some work to see if some product of natural numbers will generate this number. In fact, this number is the product of 113 and 127. So, it does not belong to the set of prime numbers.

The Mandelbrot Set

In the late 1970s and early 1980s, mathematicians began using computers to analyze solutions in a branch of mathematics called *complex dynamics*. This field involves sets of numbers that are defined by functions that involve *iteration* — that is, doing a calculation repeatedly. For example, take the number 1 and double it. Now double the result. Then double that result, and so on forever. This iteration will generate the sequence of numbers 1, 2, 4, 8, 16, 32, 64, 128 . . . and so on. We might represent this expression as $2z \rightarrow z$, meaning that we multiply the number (z) by 2, and this becomes the next value of z. This particular sequence is *unbound*, meaning the numbers just get larger and larger without limit.

Now, let's try a different iteration. Take the number 1 and divide it by 2. Then *divide* the result by 2, and that result by 2 and so on. This iteration produces the numbers 1, ½, ¼, ⅛, 1/16, 1/32, and so on, getting closer to but never actually reaching zero. We represent this iteration as $z/2 \rightarrow z$. This iteration is *bound*, meaning the numbers never exceed a certain value (in this case, 1). So *unbound* sequences get larger without limit, but *bound* sequences have a largest number that none of the members will exceed.

With the iterations mentioned above, it is easy to see whether they are bound or unbound. But with other iterations, it is not so obvious. Some iterations must be done many times before we know whether the sequence is bound or unbound. In the late 1970s and early 1980s, computers were finally fast enough and affordable enough to be useful in this kind of analysis. This allowed mathematicians to explore sets that are defined by iterative functions. One set in particular that caught their interest came to be called the Mandelbrot set, after Benoit Mandelbrot, who explored and popularized this particular number set.

The Mandelbrot set involves the iteration $z^2 + c \rightarrow z$, where z is initially zero. This means that the value of the number z is squared and then added to a different number (c) to become the new value of z, which is plugged back into the formula and so on.

The images in this book therefore represent an infinitesimal glimpse into the mind of God.

The Mandelbrot set is defined as the set of all numbers c for which the sequence z remains *bound* according to this iteration. A more conventional way to write this is:

$$z_n^2 + c = z_{n+1}$$

Mathematicians use subscripts to indicate that the next value of z (which is z_{n+1}) depends on the *current* value of z (which is z_n). The number "c" is the number we are testing to see if it belongs to the Mandelbrot set. So if you want to see if the number 1 belongs to the set, then c = 1. The symbol z_n represents a sequence of numbers. So the sequence z_n might look like this: 0, 1, 2, 5, 26, 677.... That's an unbound sequence because it gets larger without limit. The subscript n represents the position in the sequence starting with n = zero. So $z_0 = 0$, and $z_4 = 26$. By construction, the first element in the sequence is always zero. So $z_0 = 0$ always. After that, each value of z depends on the previous value of z (and the value of c), according to the formula above. So, the symbol z_{n+1} means that the *next* number in the sequence will be the square of the previous number (z_n^2) plus c.

The easiest way to get a feel for this is to simply do a few examples. We ask, "Is the number 1 part of the Mandelbrot set?" since we are evaluating the number 1, c = 1. So our iterative formula will be:

$$z_n^2 + 1 = z_{n+1}$$

By definition, the first value in the sequence of z_n is always 0, so $z_0 = 0$. Starting with z = 0, we take this number (zero), square it (it's still zero), and add it to c (which is one) to get the next z in the sequence. Namely, $0^2 + 1 = 1$. So the next value in our sequence of z_n is 1. Now we do this again. Take 1, square it (it's still 1), add it to 1, and this results in the number 2. Do this again: 2 squared (is 4) plus one is 5. Do this again: 5 squared (is 25) plus one is 26. So our sequence of z_n looks like this: 0, 1, 2, 5, 26, 677.... We can see that this sequence is unbound. Since the number 1 (c = 1) generated a sequence of z_n that is *un*bound, the number one is *not* part of the Mandelbrot set. We might construct a table to record this:

Number (c)	Part of Mandelbrot set?
1	No

What about the number zero? Does it belong? To test this, we set c = 0, start with $z_n = 0$, and plug it into the formula: $0^2 + 0 = 0$. So the next value of z is also 0. Doing this again, we see the next value of z is also 0 and so on. Our sequence of z_n is: 0, 0, 0, 0, 0, 0, 0. . . . Now this sequence is clearly bound because the value of z will never exceed zero. Since the sequence is bound, the number we were checking, namely zero (z = 0), is indeed part of the Mandelbrot set. So we can add it to our table:

Number (c)	Part of Mandelbrot set?
1	No
0	Yes

One more example involves a number that generates a very interesting sequence. Does the number *negative one* belong to the Mandelbrot set? In this case, c = -1 and substituting this into the formula we have:

$$z_n^2 - 1 = z_{n+1}$$

As before, our first value (z_0) will be zero by definition. We square this number (still zero) and subtract 1 to get -1. We then take this new value of z (negative one) and plug it back into the formula. Negative one squared (which is positive one) minus one is zero. Plugging this back in, we then get the next value of z as -1. So our sequence of z_n is: 0, -1, 0, -1, 0, -1, 0, -1. . . . This sequence cycles between two values forever! But clearly the sequence is bound. Its absolute magnitude never exceeds 1. Therefore, the number -1 is indeed part of the Mandelbrot set, and we can add it to the table:

Number (c)	Part of Mandelbrot set?
1	No
0	Yes
-1	Yes

You can see why this branch of mathematics flourished after the development of computers. It is tedious to do these computations by hand. But computers can do such tasks quickly and test many different numbers to see if they belong to the Mandelbrot set.

Complex Numbers

There is one more nuance to the Mandelbrot set before we get to the really interesting stuff. The Mandelbrot is not limited to the so-called "real numbers," but also includes complex numbers and "imaginary numbers." I hate the terminology because it is misleading. The name suggests that imaginary numbers are made-up or somehow less valid than the so-called "real" numbers. But in fact, both real numbers and imaginary numbers do exist. They are equally legitimate and are useful. And the terminology has become standard.

An imaginary number is a number that when squared produces a *negative* number. So imaginary numbers are not positive (because a positive number squared is a positive number), and imaginary numbers are not negative (because a negative number squared is a positive number), and imaginary numbers are not zero (because zero squared is zero). So how can you have a number that is not positive, not negative, and not zero?

To answer this, consider a number line. Those numbers to the right of zero are positive. Those numbers to the left are negative. We can think of the imaginary numbers as being on a different axis, directly above or below zero (see figure 1.1). Such numbers are not to the right (not positive) and not to the left (not negative) and yet are not zero. This satisfies the definition of an imaginary number.

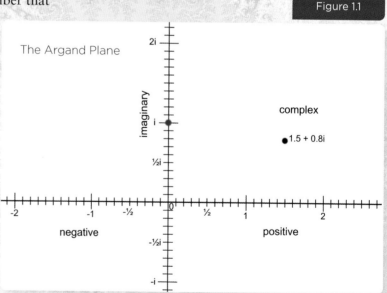

Figure 1.1

The imaginary number equivalent of the number one is symbolized by a lowercase letter i. This number squared is equal to negative one. By convention, we place i directly above zero. It is placed at the same distance as the number one is from zero. All the other imaginary numbers are generated by multiplying i by any real number. So, 2i, 3i, and so on. These numbers obey the ordinary rules of mathematics; it's just that whenever you encounter an i^2, this is equal to negative one. For example, $(3i)^2 = -9$.

We might also consider numbers that are not on either axis. These are called *complex numbers*. They get their name because they have two parts: a real part and an imaginary part. For example, the number 3 + 2i is complex. It has two parts but is one number. We can plot it using the real component as the x-coordinate and the imaginary component as the y-coordinate. This way of depicting complex numbers by coordinates on a surface is called an *Argand plane*.

> An imaginary number is a number that when squared produces a *negative* number.

The Mandelbrot set includes some complex (and imaginary) numbers as well. To see which numbers belong to the Mandelbrot set, we simply set c equal to the number in question and see what sequence of z emerges. If the sequence gets larger without limit, then c is not part of the Mandelbrot set. But if the sequence of z remains bound, then c is part of the Mandelbrot set. Computers can do these calculations quickly for many numbers. In the 1980s, mathematicians and programmers began to use computers to make a map of the Mandelbrot set. When they did this, a remarkable pattern emerged.

Mapping the Mandelbrot Set

Using computers, we can make a map of the Mandelbrot set in the Argand plane. Recall that each point on the plane represents one complex number, with the x-coordinate representing the real component and the y-coordinate representing the imaginary component of that number. The computer quickly checks each number by running it through the formula, generating a sequence of z_n, which either remains bound (for numbers belonging to the Mandelbrot set) or becomes larger without limit (for numbers *not* in the Mandelbrot set).

But how does the computer decide if the sequence grows larger *forever* or remains small forever? The computer can do a lot of iterations quickly, but it cannot do them forever! We humans do not need to run the iteration forever to see what will happen because we intuitively understand patterns. We know that the sequence 1, 2, 4, 8, 16, 32 . . . will get larger without limit because we can see that each number is twice the previous number. We understand that the pattern 0, -1, 0, -1, 0, -1 will be bound because it cycles. But computers have no such comprehension or intuition.

Since computers have no real understanding, in practice, the way they decide if a sequence is bound is for their programmer to set an "escape value" and an "iteration limit." In other words, if after 1,000 iterations the value of z is still smaller than, say, 10, we can be pretty confident (though not absolutely sure) that the sequence is bound. If, however, the value of z is larger than 10 after only 4 iterations, we can be confident that the sequence will grow without limit. For the Mandelbrot set, mathematicians have shown that the escape value can be as small as the number 2. That is, for a given c, if any number in the sequence z_n is larger than 2, then the sequence will grow large without limit and is unbound. So most programmers set the escape value to the number 2. The iteration limit is harder to guess, but suffice it to say that larger values produce a more accurate map.

So the computer systematically checks

> Using computers, we can make a map of the Mandelbrot set in the Argand plane.

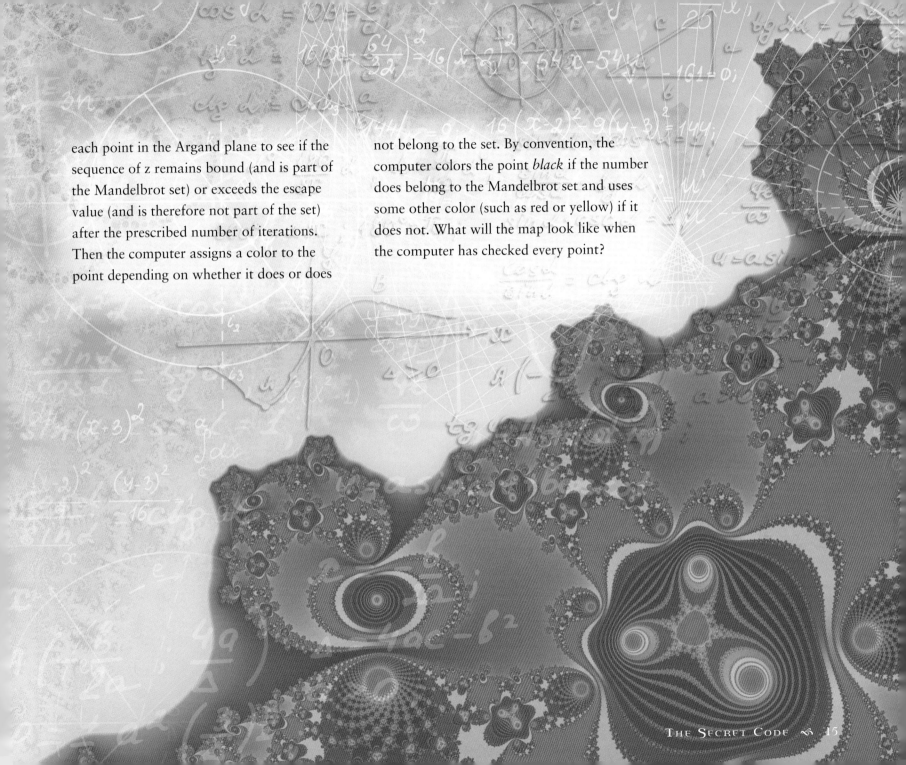

each point in the Argand plane to see if the sequence of z remains bound (and is part of the Mandelbrot set) or exceeds the escape value (and is therefore not part of the set) after the prescribed number of iterations. Then the computer assigns a color to the point depending on whether it does or does not belong to the set. By convention, the computer colors the point *black* if the number does belong to the Mandelbrot set and uses some other color (such as red or yellow) if it does not. What will the map look like when the computer has checked every point?

Now the Really Cool Part

Naïvely, we might think the map would be a circle or some basic shape based on the simplicity of the definition of the set. After all, $z^2 + c$ is a pretty simple expression. But instead, the map of the Mandelbrot set turns out to be remarkably interesting and complex, as shown in figure 1.2. This basic shape was first discovered and plotted (in black and white) in 1978, but at much lower resolution than we can do today.

In the 1980s, Benoit Mandelbrot developed software to improve the plotting of the Mandelbrot set and its exterior, eventually in shades of color that represent how quickly the sequence z_n grows — how many iterations it took to exceed the escape value. In other words, for numbers that do not belong to the Mandelbrot set, the sequence of z_n can grow large rapidly (such as $z_n = 0, 10, 1000, 100000, 1000000000$), in which case the second iteration has already exceeded the escape value of 2. Or, the sequence can grow slowly (such as $z_n = 0, 1, 1.1, 1.15, 1.19$), in which case it might take a hundred iterations or more for the sequence to exceed the escape value. By convention, we normally use bright colors (like bright yellow) for numbers where the sequence of z_n grows slowly, indicating numbers that are *very close to* (yet not part of) the Mandelbrot set. And we use darker colors (like deep red) for numbers where the sequence of z_n grows quickly, indicating numbers that are far away from being part of the Mandelbrot set. And again, numbers that are part of the Mandelbrot set are colored black.

So, in figure 1.2, the black regions represent all complex numbers that belong to the Mandelbrot set. The yellow regions represent numbers that are very close to the set, but do not belong to it. And the dark red regions represent numbers that are not even close to being part of the Mandelbrot set. Using this map, we can easily check whether any given number belongs to the set simply by checking the color at its coordinates because the computer has already done the calculation. We can see that -1.5 does belong to the Mandelbrot set, but +1.5 does not. We can see that zero belongs (as we proved earlier), but 2i does not. And so on.

Now the amazing thing here is not so much that we have a convenient map, but rather the shape of the map itself. No one had imagined that this map of the Mandelbrot set would have such an amazing and complicated shape. And as we will see later, when we zoom in, some sections of the Mandelbrot set are immensely beautiful. The shape itself has wonderful mathematical properties. I suppose that is not too surprising given that it is a mathematical graph. But the particular geometric and mathematical properties it exhibits were a surprise to everyone. Who knew that such properties had been hidden in the little formula $z^2 + c$?

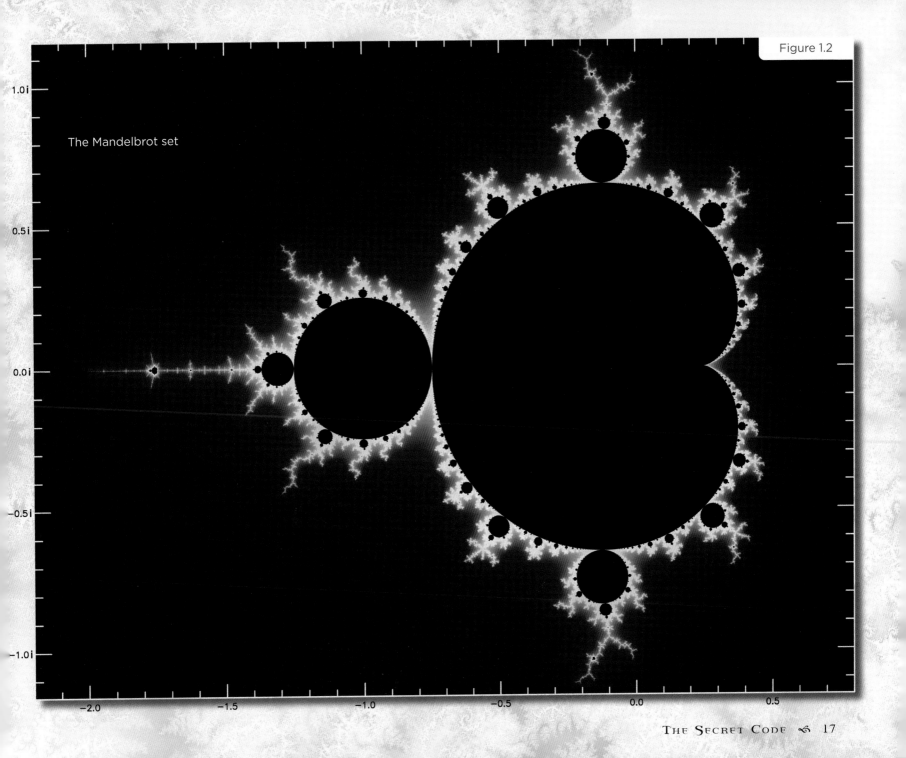

Figure 1.2

The Mandelbrot set

Geometry

At first glance, we notice that the Mandelbrot set has three types of geometric shapes. The largest and most prominent is the heart-shaped structure. This shape is called a *cardioid*. It is the shape generated when you roll one circle around another of equal size, keeping your pencil affixed to a point on the rolling circle. The cusp of the cardioid is located exactly at ¼, and its opposite side ends at exactly -3/4. The cardioid has an area of $3\pi/8$.

Next, we notice lots of perfect circles budding off the cardioid. The largest of these circles is affixed to the left side of the cardioid, is centered exactly on the number -1, and has a radius of exactly ¼. Another smaller circle grows off of its left side, with another growing off of it, and so on, as far as the eye can detect. The second largest circles in this map are affixed to the top and bottom of the cardioid.

Cardioid - sinusoidal spiral - mathematical plane curve.

Next, we notice lots of perfect circles budding off the cardioid.

The third shape we notice are thousands of tiny "branches" or "dendrites" that are rooted in the circles that bud off of the cardioid. All of these dendrites are very "wiggly" with one exception: the antenna extending directly to the left on the real number line is perfectly straight and ends at exactly c = -2. It may seem at first that these branches are not part of the Mandelbrot set because most of them are not colored black. But since we see bright yellow in these branches, we must conclude that such points are extremely close to the Mandelbrot set. In other words, the actual (black) threads are too thin to be visible but are surrounded by yellow. They branch into dendrites, which then branch into more dendrites. We will see that this type of feature is common in the Mandelbrot set.

The Mandelbrot set has an infinite number of smaller versions of itself built into itself!

Thousands of tiny "branches" or "dendrites" that are rooted in the circles that bud off of the cardioid

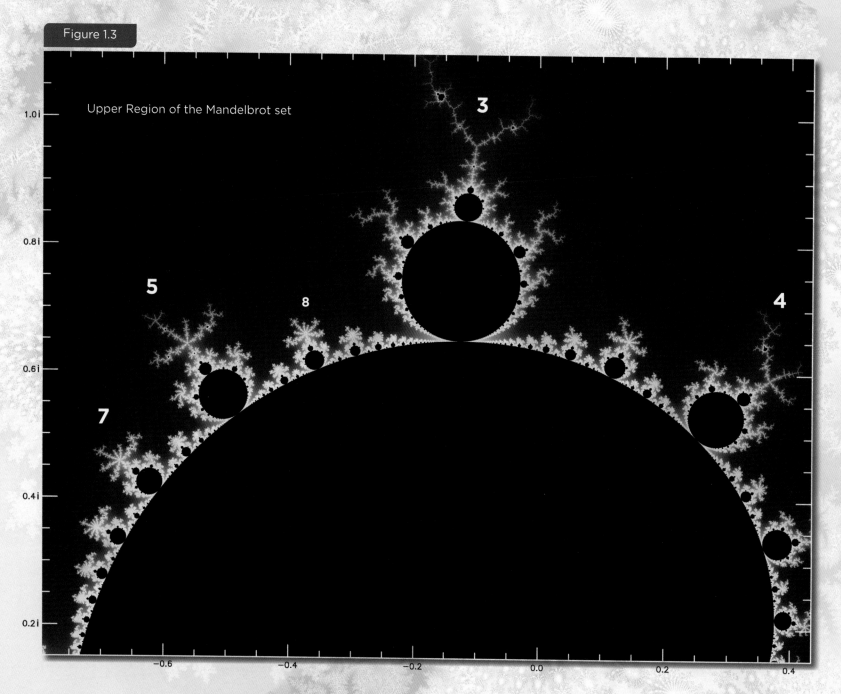

Figure 1.3

The Branches are Smart!

Let us explore the dendrites near the top of the Mandelbrot map. We can zoom in on this area by having the computer check these values at finer resolution than we did previously. The resulting map (figure 1.3) shows that the branches rooted in each circle exhibit some fascinating properties. The branch growing above the largest circle splits into two more, for a total of three that meet at an intersection. The next largest circle to the left has a branch that splits into a total of five. The next largest circle to the left branches into 7, the next 9, then 11, 13, and so on. These branches cover all the odd numbers in perfect sequence to infinity.

Just to the right of the largest circle, on the next largest circle, we count four intersecting branches. The next largest circle to the right has branches that split five ways, the next 6, 7, 8, etc. So the circles on the right side have branches that count all the numbers, both even and odd, from three to infinity! It seems that the branches of the Mandelbrot set know how to count. But wait — there's more!

Consider the largest circle in figure 1.3 (that has three branches) and the next largest to its left (which has five branches). Now examine the largest circle that is between them. It branches into 8 parts. Why is that significant? Eight is three plus five — the sum of the branches on the surrounding circles. In fact, this is the case for all the circles! Each circle that is the largest in between two larger ones has the sum of their branches. It seems that these branches not only know how to count to infinity, but they can also add!

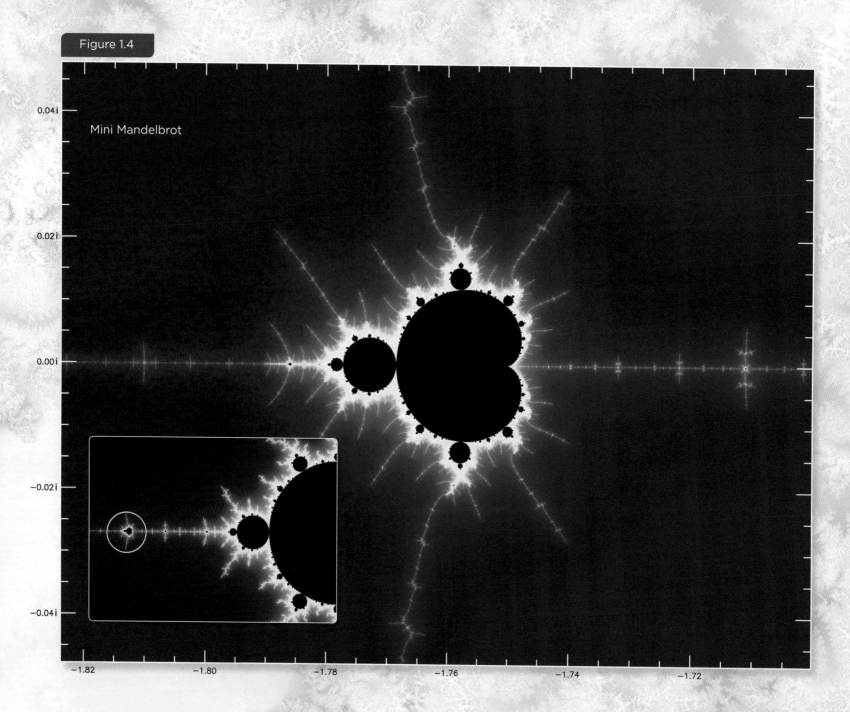

Figure 1.4

Mini Mandelbrot

Mini and Mini-Minis

Returning to figure 1.2, let's now examine one of the most fascinating properties of the Mandelbrot set. Consider the straight, long spike on the left of the image — the only non-wiggly dendrite. About two-thirds of the way to the left is a "bump," with tiny branches extending above and below. When we zoom in on this shape, we find in figure 1.4 that it is a tiny version of the entire Mandelbrot set! This mini-Mandelbrot is nearly identical to the original. It has the large cardioid with circles budding off of it, the largest circle is on the left, and a spike is extending to the left.

But there are slight differences. When we compare figures 1.2 and 1.4, we see that this baby Mandelbrot has extra spikes extending away from it. We zoomed in on the spike of the (large) Mandelbrot and found that the mini version has extra spikes. Can this be a coincidence? Second, we note that this mini Mandelbrot is growing off of another and therefore has a spike entering the cusp of the cardioid. The large Mandelbrot set lacks this trait because it does not stem from a larger structure.

Of course, the mini Mandelbrot also has a spike on its left, just like the original large version. And this spike also has a small bump on it. Zooming in on this tiny bump, we find that it too is a tiny version of the entire Mandelbrot set (figure 1.5). It is a "mini-mini-Mandelbrot"! At this scale, we find it useful to employ more complex color schemes to reveal intricate details. In this case, as the iterations increase, the palette goes from red, to yellow, to light blue, to white.

Again, the main features are identical to the entire Mandelbrot set — the cardioid, the circles, and the branches. But since we zoomed in on the spike of the mini, which was on the spike of the original, this mini-mini-Mandelbrot has extra spikes extending away from it. Apparently, miniature versions of the Mandelbrot set inherit the geometric characteristics of the part of the parent from which they extend. This mini-mini-Mandelbrot also has a spike on its left, which has a bump. Of course, this turns out to be an even smaller version of the Mandelbrot map (figure 1.6). Figures 1.5 and 1.6 were plotted with the same color palette for easy comparison.

The two maps are nearly identical apart from size, and the mini-mini-mini has gained extra spikes that stem off the other spikes. This mini-mini-mini also has an even smaller version budding off of its tail (figure 1.7), which also has an even smaller version budding off of its tail (figure 1.8), and so on. This pattern apparently continues forever, with each smaller version gaining additional spikes and complexity. The Mandelbrot set has an infinite number of smaller versions of itself, built into itself! This type of structure is called a *fractal*.

Figure 1.5 — Mini Mini

Figure 1.6

Mini Mini Mini

A fractal is any geometric shape that has parts that resemble the whole. If we adjusted the contrast on the "mini-Mandelbrots" in figures 1.2 through 1.8 so that the exterior spikes were not visible, they would be virtually indistinguishable from the entire Mandelbrot set. We would not know if we are viewing the entire map or if we have zoomed in by a factor of a billion or a hundred quadrillion. We see the same basic type of shape no matter how much we zoom in. This property of fractals is called *scale-invariance*. In the next chapter we will see that the Mandelbrot set has many sections that exhibit scale-invariance.

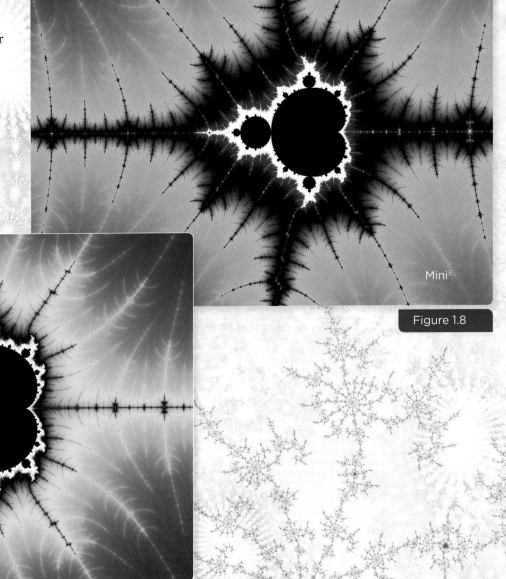

Figure 1.7

Figure 1.8

Figure 2.1

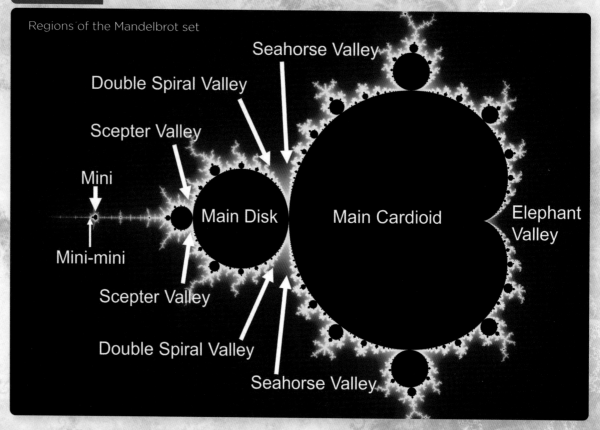

Regions of the Mandelbrot set

People have named various sections of the map of the Mandelbrot set (see figure 2.1). Some of these sections turn out to be immensely beautiful. We begin our journey with Seahorse Valley. This section is found on the right side of the valley between the cardioid and the largest circle. It doesn't matter whether we select the upper valley or the lower one. They are perfectly symmetric mirror images.

Seahorses in the Mandelbrot Set

Recall that the circles near the top of the cardioid have branches that split into 3, 5, 7, 9, 11…increasing by 2 with each circle as we follow them down into the valley. As the branches increase, they begin to form a complex web structure that resembles a seahorse. If we choose the lower valley, the seahorses will appear right-side up — see figure 2.2. Each seahorse has two extra branches from the one that came before it, and the branches approach infinity as we approach the cusp of the valley. Figure 2.3 shows a close-up of some of the seahorse structures deep in the valley. Seahorses far from the cusp have fewer branches and appear more fragmented, as in figure 2.4. As we go deeper into the valley, the seahorses gain branches, become smoother, and their tails become increasingly spiraled as shown in figure 2.5.

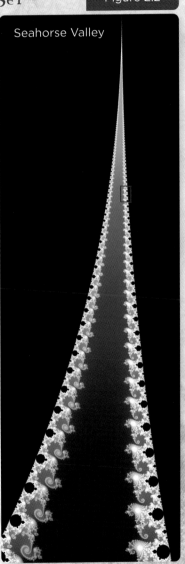

Figure 2.2

Seahorse Valley

Figure 2.3

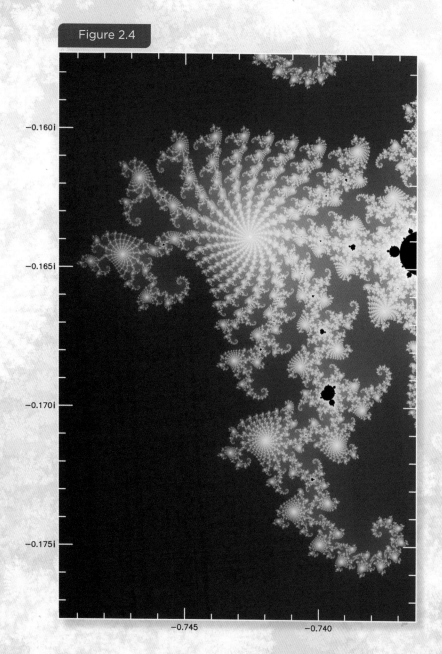

Figure 2.4

The color scheme we choose for these plots is arbitrary. We can use blues, reds, greens, or whatever we want for those points outside the Mandelbrot set. Interior points are always black by convention. But the structures are best visualized when we use brighter colors to represent points that are very close to the Mandelbrot set. The bright colors seen in figure 2.3, therefore, must be very close to a black thread, even though the thread itself may be too thin to be visible. Hence, the spiral and cobweb-like structures seen in the seahorses do indeed belong to the Mandelbrot set, even though the branches are often too thin for the black to be visible (or printable).

The Mandelbrotset is a *connected* set. We can think of this geographically. If the Mandelbrot set represented a landmass, and its exterior represented the ocean, then the Mandelbrot set would have many peninsulas but no islands. It would be impossible to boat completely around any part of the Mandelbrot set without boating around the entire Mandelbrot "continent."

Figure 2.5

The fact that we are dealing with a connected set means the branch extending away from any circle in figure 2.3 is so wiggly and so complex that it weaves in and out and branches into other branches, all of which form the complex seahorse structure that we see. Think of how complicated that shape must be! In fact, the branches are "infinitely wiggly." The exterior of the Mandelbrot set is as wiggly as is possible in a two-dimensional plane. In other words, if the Mandelbrot perimeter were any wigglier, it would pop out of the Argand plane into the third dimension.

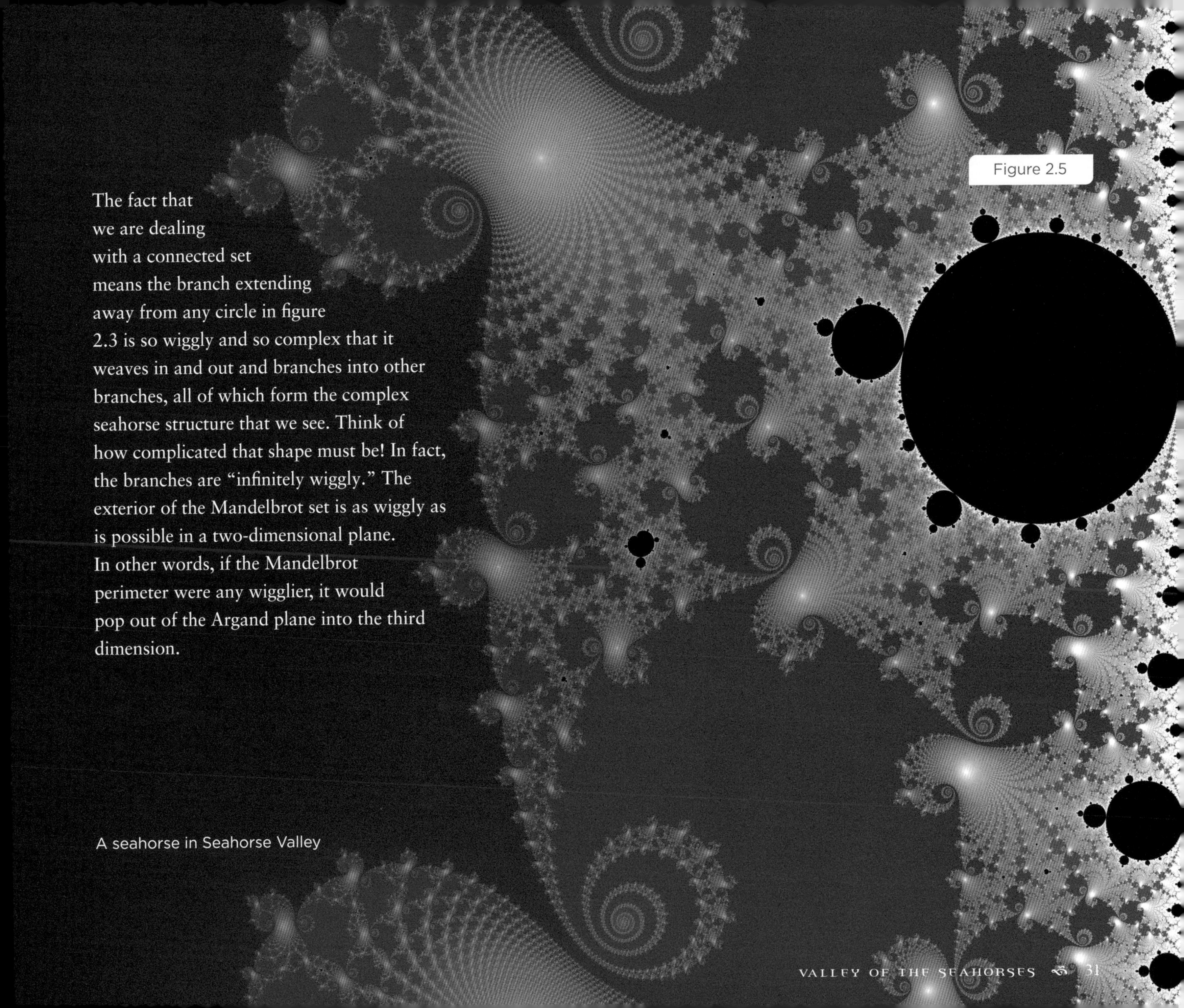

A seahorse in Seahorse Valley

Figure 2.6

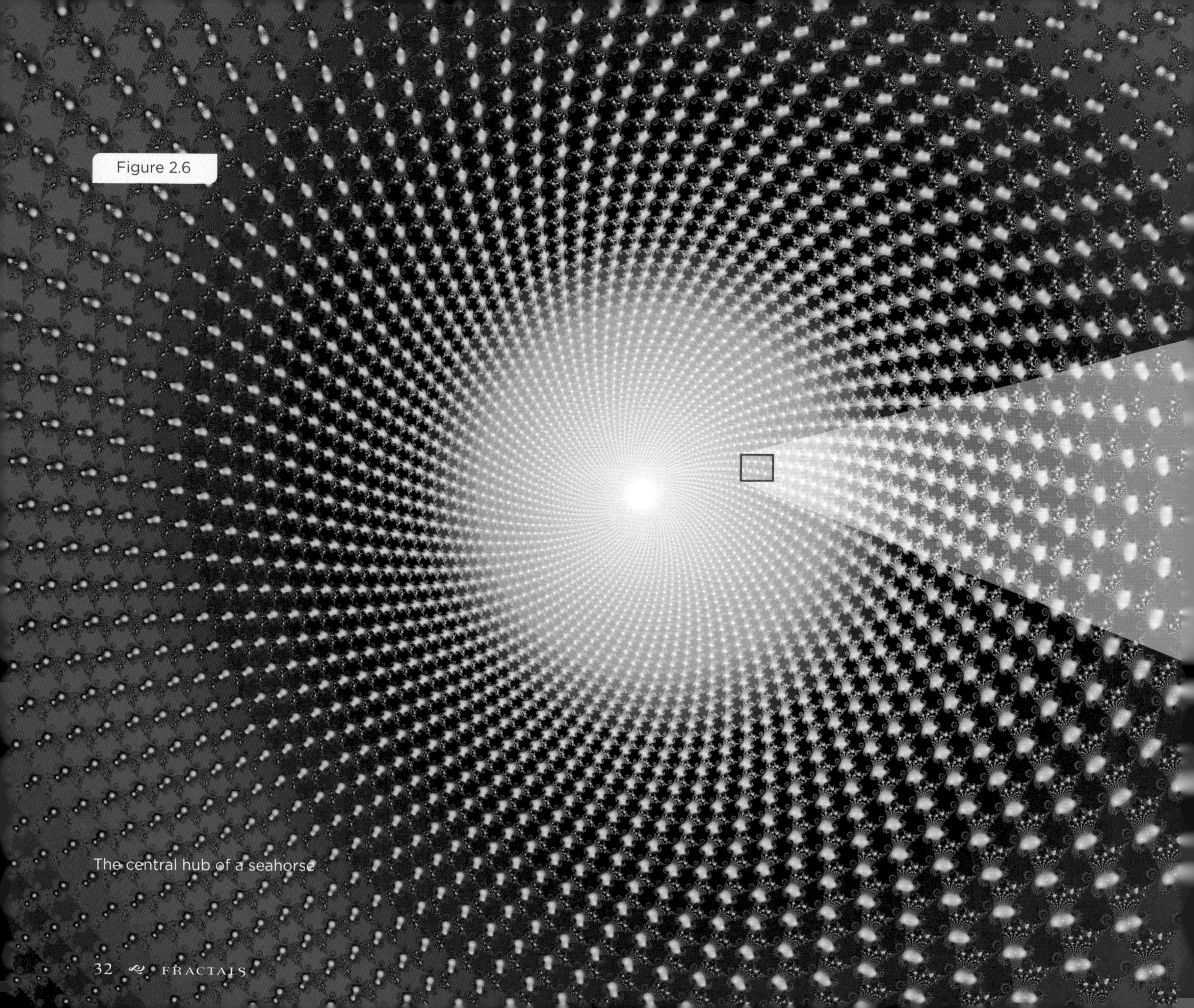

The central hub of a seahorse

32 ~ FRACTALS

Figure 2.5 is a higher-resolution plot of the upper seahorse in figure 2.3, and it allows us to see the structure in greater detail. The tail forms a beautiful spiral, which continues inward infinitely. The upper body of the seahorse resembles a dense spiderweb, with spokes extending away from a bright core. The head of the seahorse also curls into a spiral at the bottom. Notice that the "face" of this seahorse is itself an entire smaller seahorse, with a face that is also a smaller seahorse, and so on forever. Tiny seahorses can also be found throughout the body of the larger seahorse. The scale-invariance of this fractal is remarkable.

As we zoom in on the central spiderweb structure, we begin to get a sense of the complexity of this shape (see figure 2.6). In fact, we could zoom in on the center of this structure forever, and it never ends. So we instead examine a section of the strands (red box) in higher resolution (figure 2.7). Amazingly, we see that these strands are themselves made of spiderwebs and spirals on a smaller scale.

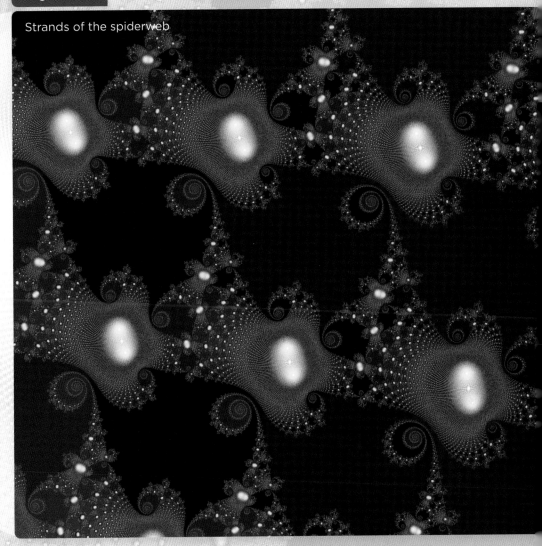

Figure 2.7

Strands of the spiderweb

Figure 2.8

These are the basic structures seen in a seahorse, and we now see them here as well. This is a further illustration of the scale-invariance seen in fractals. On this scale, we see that each spiderweb structure has not one, but two bright yellow cores, which come in pairs. Zooming in on the lower central pair of cores, we obtain figure 2.8.

In figure 2.8, we see the two bright central hubs of the spider webs near the top and bottom. Midway between them, we see a square structure formed by four bright hubs. In zooming in, the hubs have gone from two to four. As we zoom in on these four hubs, we will find that they surround eight hubs, which surround sixteen hubs, and so on — all the powers of 2. This is evidenced in figure 2.9 and figure 2.10.

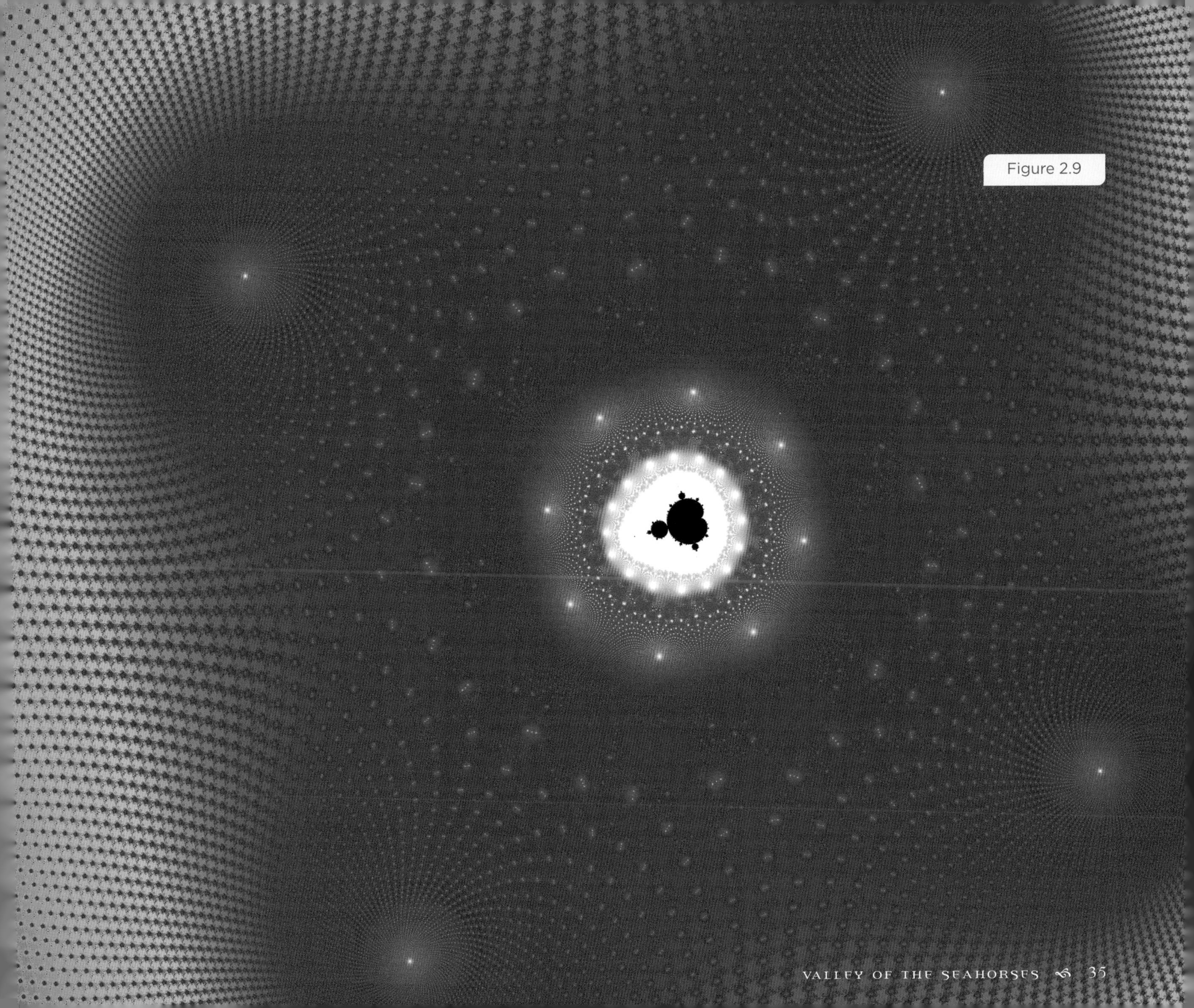

Figure 2.9

VALLEY OF THE SEAHORSES 35

In the center of figure 2.9, we see another miniature copy of the entire Mandelbrot set! Here we can clearly see the four bright hubs on the perimeter and eight hubs surrounding the mini. Zooming in further (figure 2.10), we see the 8 hubs on the perimeter, which surround 16 bright hubs, then 32, 64, 128, etc., as we approach the main body of the mini Mandelbrot. These are all the powers of 2. So not only does the Mandelbrot map know how to count and add as we saw in the branches, but it also understands powers of 2. Why powers of 2? Could this be related to our iterative formula ($z^2 + c$)?

We again see that the mini Mandelbrot here resembles the entire Mandelbrot in terms of its basic shape. But around the exterior we find incredibly complex and beautiful spiderweb structures. So the exterior of the mini resembles the section of the parent on which it grows. And that's just the beginning.

> So not only does the Mandelbrot map know how to count and add as we saw in the branches, but it also understands powers of 2.

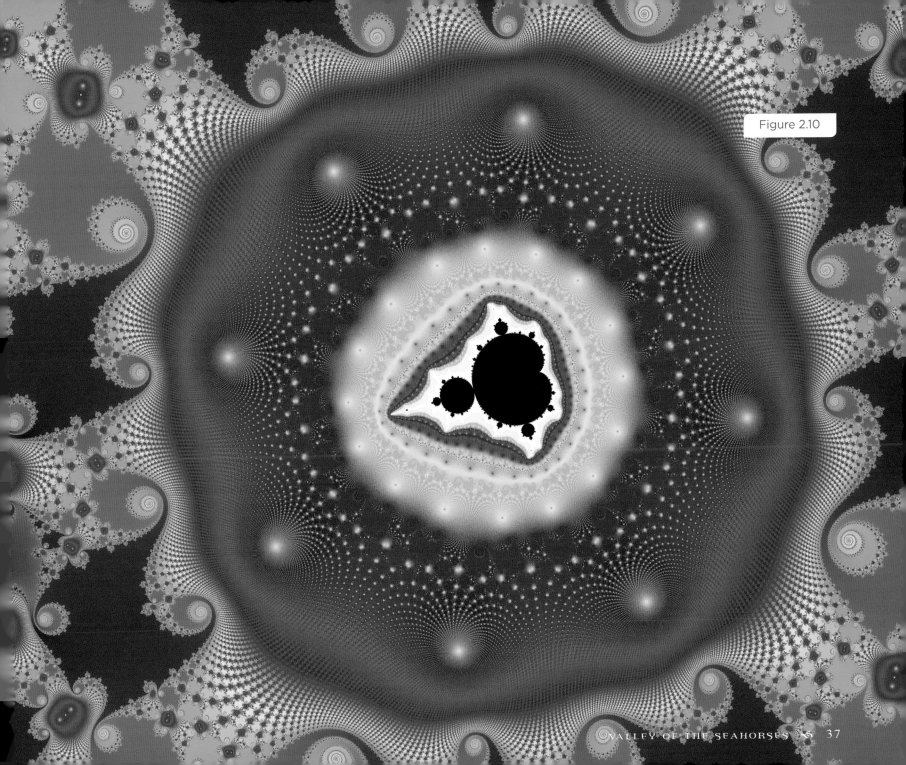

Figure 2.10

Valley of the Double Spirals

One of the most beautiful areas of the Mandelbrot set occurs on the left side of the cusp between the main cardioid and the main disk. This is the same cusp as the seahorses. But whereas the right side of the valley has geometry that resembles seahorses, the left side has galaxy-like spirals and is called Double Spiral Valley.

As before, the valley below the x-axis is a mirror image of the valley above. So we will select the upper valley this time.

A close inspection of the left side of this valley reveals a sequence of double spirals, each one branching from a disk (see figure 3.2). But what do we mean by a *double spiral*? Figure 3.1 illustrates

> The double spiral has two non-intersecting strands that wrap around each other.

Figure 3.1 — A double spiral and single spiral

the difference between a double spiral (left) and a single spiral (right). The double spiral has two non-intersecting strands (shown in blue and red, respectively) that wrap around each other. The single spiral shown in green has only a single strand. The spirals we saw in Seahorse Valley were single.

Near the top of the valley, far from the cusp, the double spirals are rather *open* (see figure 3.3 on following page). In other words, as we go from the periphery of the spirals toward the core, the spiral winding is slight. Farther down the valley, the spirals wind increasingly tighter and become pinched by the surrounding structures (figure 3.4). In figure 3.2, we are at an intermediate depth where the double spirals resemble the shape of spiral galaxies.

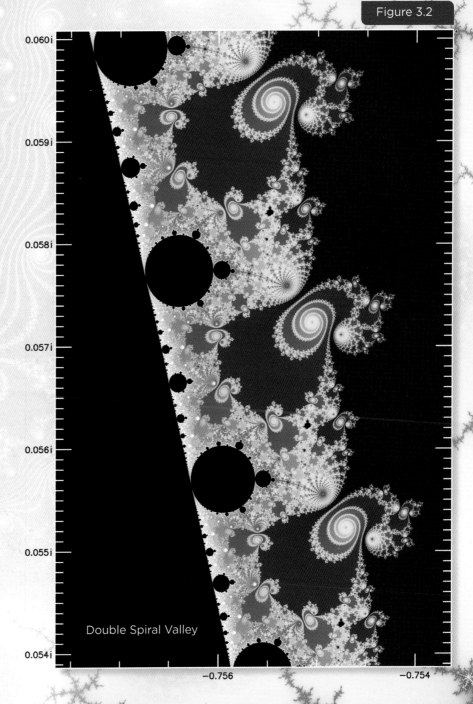

Figure 3.2

Double Spiral Valley

Figure 3.3

Figure 3.4

VALLEY OF THE DOUBLE SPIRALS

Figure 3.5

Figure 3.5 is a high-resolution plot of the central double spiral seen in figure 3.2. The fractal nature of this structure is easy to see. The basic structure is made of many smaller copies of itself. Notice, for example, that one strand of the double spiral branches off into a spiderweb structure on the right side. The upper right edge of this spiderweb contains another much-smaller double spiral with a spiderweb extending from its right side, and so on, forever.

Zooming in on this spiral (figure 3.6) we see that it is, in fact, a double spiral. That is, it has two independent strands that wrap around each other. You can check this by tracing one of the strands all the way around until it comes back

Figure 3.6

VALLEY OF THE DOUBLE SPIRALS ❧ 45

Figure 3.7

to the same angle and confirming that there is a strand between the starting point and ending point. The spirals found in Seahorse Valley were single spirals — one strand that wraps around itself. This double-spiral structure is similar to a grand design spiral galaxy. However, we could zoom in on the center of this spiral forever. It continues to spiral infinitely! So, instead we will zoom in on one of the strands (in the rectangle), and the result is figure 3.7. We see that the strands are themselves composed of double spirals — two large ones are visible near the top. And we also see the spiderweb structure we saw in the Valley of the Seahorses. But there is an additional structure that vaguely resembles a bowtie. There are three large "bowties" that stretch across the center of the image, and many smaller ones are visible throughout.

Figure 3.8

Zooming in on the bowtie just left of center is figure 3.8. We see that it consists of 2 double spirals that cross in the middle, forming an "X" shape. Zooming in on the center of this "X" (figure 3.9), we see that it consists of 4 double spirals, surrounding 8 double spirals, and so on, all surrounding another mini Mandelbrot. Zooming in again (figure 3.10) reveals a closer look at the 8 double spirals surrounding the mini and the 16 double spirals farther in.

Again, we see that the miniature version inherits the characteristics of the portion of the parent on which it is found. Namely, we zoomed in on a double spiral, and we found that the mini Mandelbrot is surrounded by an elaborate and beautiful array of double spirals. Of course, this miniature Mandelbrot has a spike extending away from the largest circle on the right-hand side, and the small bump is revealed (figure 3.11) to be yet another even smaller miniature Mandelbrot. But unlike the mini budding off of the spike of the main (large) Mandelbrot, this mini has an intricate tapestry of double spirals extending away from it.

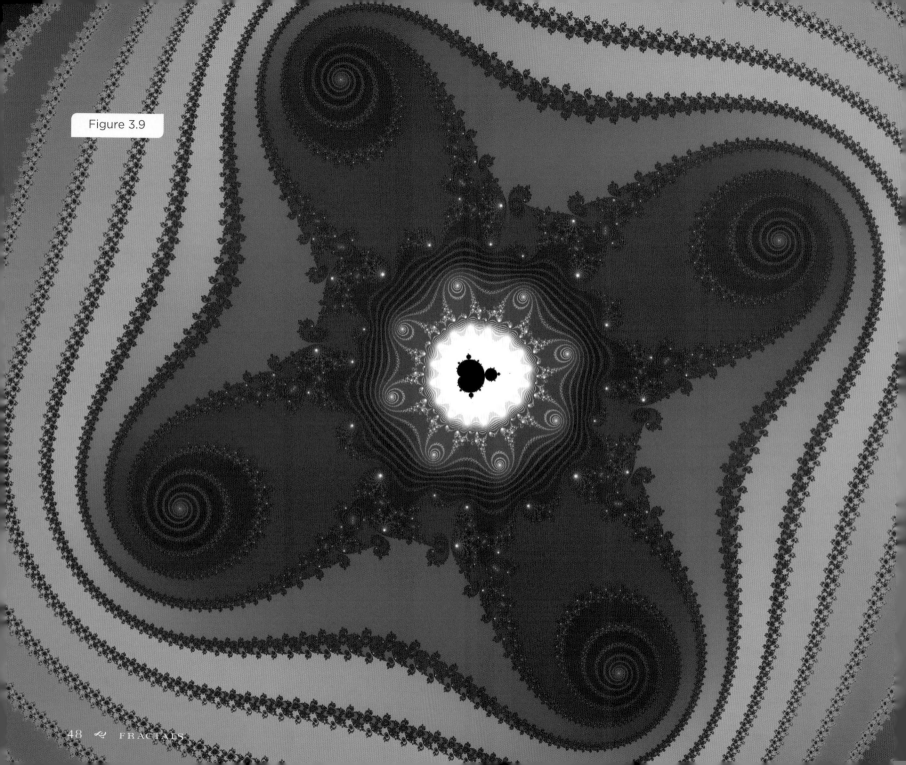
Figure 3.9

Figure 3.10

VALLEY OF THE DOUBLE SPIRALS

Figure 3.11

A Glimpse into Mind of God

Now consider this. The Mandelbrot set apparently contains an infinite number of smaller versions of itself. After all, the miniature Mandelbrot seen in figure 3.11 also has a Seahorse Valley and a Double Spiral Valley. And when we zoom in on these, we will find even smaller miniature Mandelbrots with their own valleys, and so on, forever. But unlike manmade structures, the Mandelbrot set becomes more complex and beautiful the more we zoom in. The miniature Mandelbrots always have elaborate and decorative structures surrounding their central area in contrast to the far exterior of the entire (large) Mandelbrot set map.

So, the miniature versions of the Mandelbrot are apparently far more complicated than the entire Mandelbrot. Yet the entire Mandelbrot contains all these miniature Mandelbrots! It is mind-boggling. Not only that, but consider that since the Mandelbrot set contains an infinite number of miniature versions, and since each of these is at least as complex as the entire Mandelbrot set, they too will have an infinite number of smaller versions built into them. So each of the infinite number of mini Mandelbrots has an infinite number of mini Mandelbrots. We are not merely exploring infinity, but an infinite number of infinities!

If that is hard for your finite mind to accept, you are not alone. But it is true nonetheless. All of these infinite number of infinities exist in the mind of God, and we have discovered a few of them with the aid of computer technology. This gives us an infinitesimal glimpse into the infinite mind of God. What a mind!

Mathematicians have explored some of the implications of infinity. It is a true concept, and yet one that our finite minds cannot fully grasp. However, by using God's laws of logic and God's laws of mathematics, we can learn things about infinity through logical deduction.

> We are not merely exploring infinity, but an infinite number of infinities!

Mathematics confirms that there are indeed different types of infinity, and some are larger than others!

Just as the Mandelbrot set contains not just one infinity, but an infinite number of infinities, so the number of real numbers is larger than the number of rational numbers. Rational numbers can be expressed as the ratio of two integers, such as 1/7. Real numbers include the rational numbers, but they also include values like π and the square root of 2 — numbers that cannot be expressed as the ratio of two finite integers. Both sets are infinite, but real numbers contain an infinite number of infinities, perhaps like the Mandelbrot set. So not only is God's mind infinite, but it is infinitely infinite. The properties of numbers stem from the way God thinks. And hence, the different levels of infinity are a reflection of the mind of God.

As we continue to explore the Mandelbrot set, we would do well to remember what it is exactly that we are exploring. This intricate shape is not found in physical nature. It cannot be seen in a telescope or microscope. It is not made of atoms. Rather, it is a mathematical graph — a plot of those numbers that are bound under the iteration $z^2 + c$. This remarkable structure was not designed by man, nor by nature. The Mandelbrot set is an aspect of the internal logic of numbers that has existed from the beginning of time. Yet we could spend a lifetime exploring the astonishing beauty of this map.

We have found areas of particular beauty in the cusp between the main cardioid and the main disk. But many other cusps exist — an infinite number of them, in fact. What other wonders will we find?

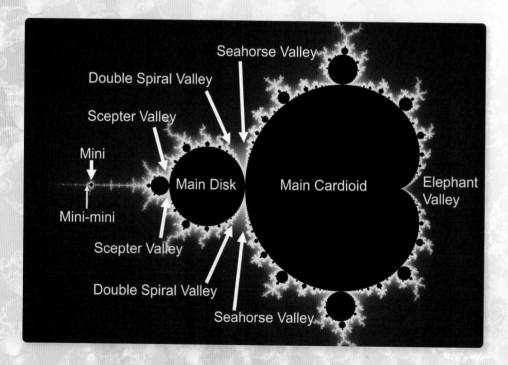

The Valley of Elephants

A close examination of the cusp on the right side of the main cardioid of the Mandelbrot set takes us to Elephant Valley (see figure 2.1 for reference). Each circle budding off of the cardioid in this area branches into a complicated shape that rather resembles an elephant (figure 4.1). Elephant Valley is the only symmetric valley in the main Mandelbrot set; the top is a mirror reflection of the bottom. And so the elephants marching along the bottom of this valley perfectly reflect those hanging upside down from the top (figure 4.2).

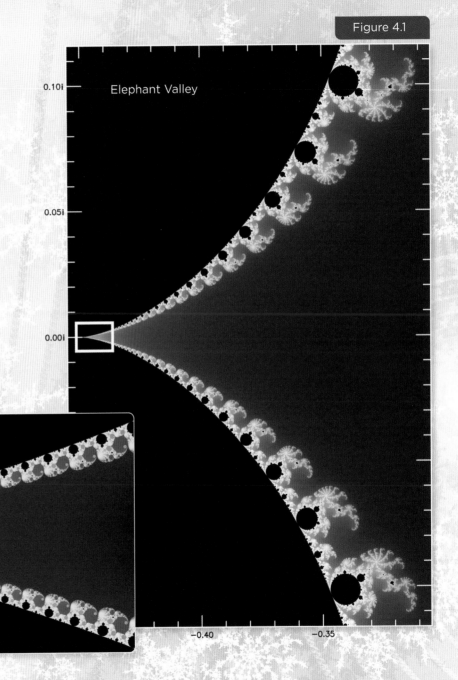

Figure 4.1

Figure 4.2

Infinite Elephants 55

Figure 4.3

The deeper we peer into Elephant Valley, the more the elephant trunks curl in on themselves. Figure 4.3 reveals two elephants deep in the valley. Note that these trunks form a single spiral — not a double. Zooming in on this spiral (figure 4.4), you can confirm by tracing this spiral around that it is a single strand. As we zoom in on the center of this spiral (figure 4.5), it goes on indefinitely; it continues to spiral to smaller scales forever.

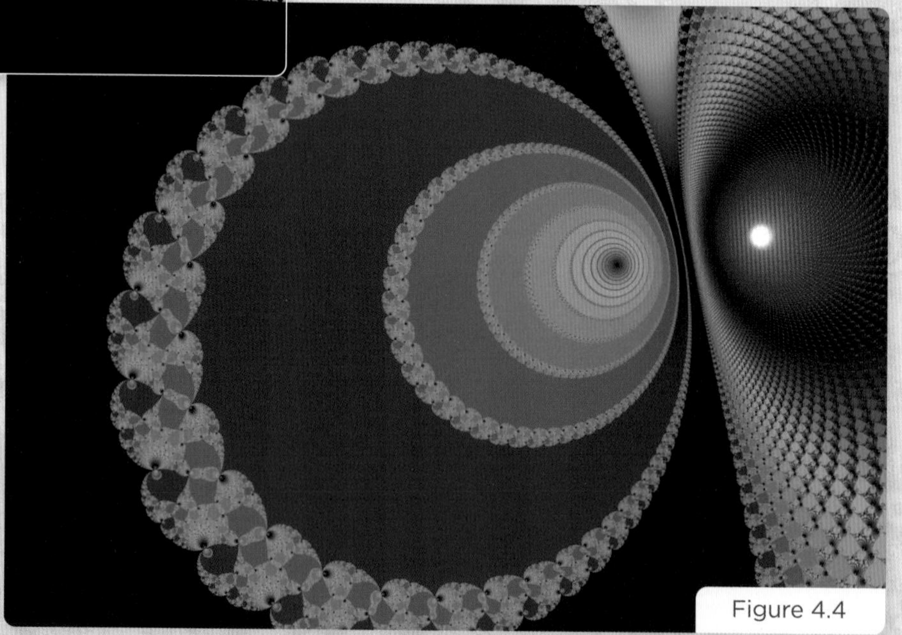

Figure 4.4

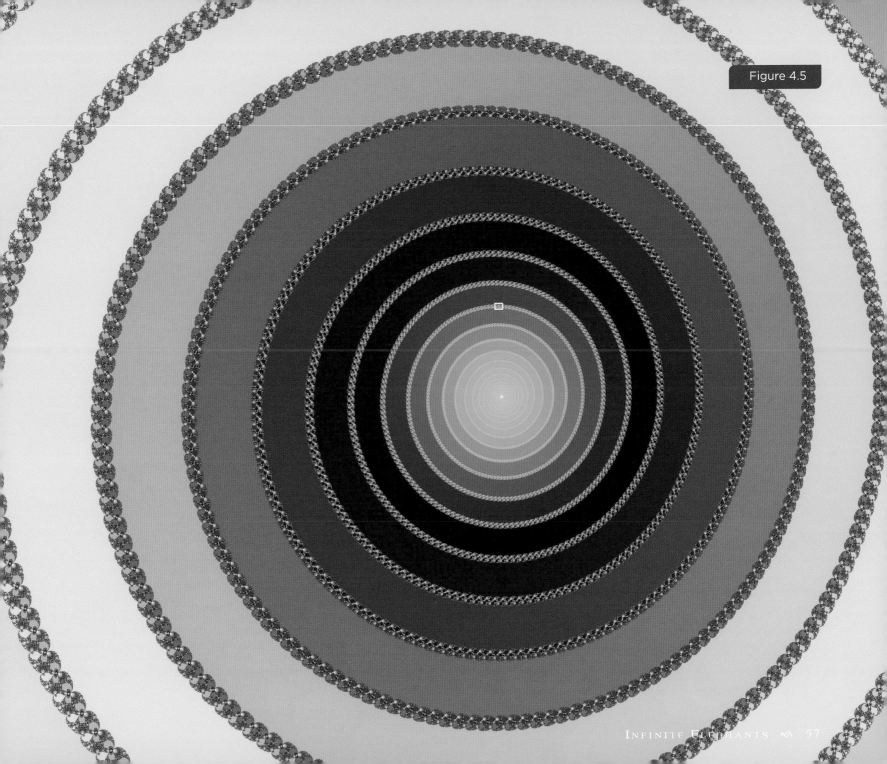

Figure 4.5

Figure 4.6

So we instead zoom in on a strand (boxed, page 57). Figure 4.6 reveals that this strand contains the spiderweb structures we have seen before. But it also contains multiple single spirals, some of which resemble elephants. Additionally, we see some double spirals, which resemble pinched ellipses.

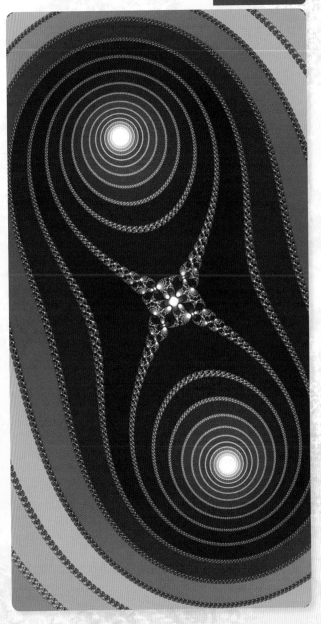

Zooming in on the ellipse near the center reveals figure 4.7 — a double spiral pinched between two spiderweb structures. In the center is a "bowtie." Figure 4.8 reveals that this bowtie is composed of two single spirals, unlike the double-spiraled bowties found in Double Spiral Valley. Yet, as before, they cross in the middle, forming an "X" consisting of four tiny spirals.

Zooming in further (figure 4.9) reveals 8 spiral hubs on the perimeter with 16 farther in and so on. The beauty here is remarkable. Who could have guessed that such delicate loveliness was built into numbers revealed by the simple formula $z^2 + c$? If indeed mathematics is a window into the mind of God, then God's thinking is not only infinite, but infinitely beautiful as well. Then again, could there be a simple secular explanation?

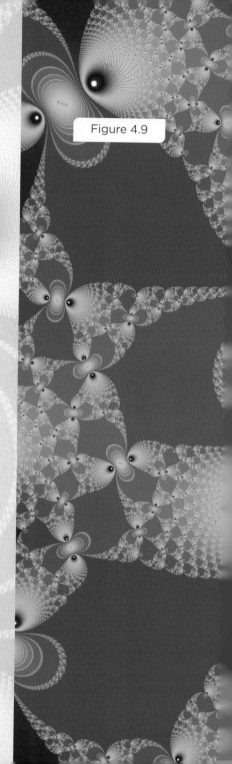

Figure 4.9

Scepters on Seahorses

There are, of course, an infinite number of valleys in the map of the Mandelbrot set — and we could spend a lifetime exploring each since every one is infinite. We will confine ourselves to just one more valley and will note that certain patterns begin to emerge. That is, some valleys resemble others but with variations. The Valley of the Scepters is located between the main disk of the Mandelbrot set and the disk immediately to its left.

As with Seahorse Valley and Double Spiral Valley, the upper Scepter Valley is a mirror image of the lower Scepter Valley. We'll select the lower valley for exploration.

The Valley of the Scepters

Figure 5.1 shows that this valley strongly resembles Seahorse Valley on the right and Double Spiral Valley on the left — just like the cusp between the main cardioid and main disk. But we see a new feature as well. Each of the seahorses now has a prominent scepter extending from its head. We might think of them as "sea unicorns." Likewise, each double spiral also has a prominent scepter extending into the gap of the valley.

Figure 5.2 is the double spiral in the left box in figure 5.1 along with its corresponding scepter. The double spiral itself is similar to that seen in Double Spiral Valley (compare with figure 3.5). However, the y-axis is reversed since we were examining the upper valley there and are examining the lower valley here. The main difference is the additional scepter on the lower right. But we also see hundreds of other smaller scepters, each rooted on a double spiral. These scepters appear nearly identical to the large one. The scale-invariance is remarkable.

Figure 5.3 is the detailed plot of the right-hand box in figure 5.1, revealing the intricate structure of the seahorse and its scepter. Again, the structure is very similar to that found in Seahorse Valley (compare with figure 2.5). The main difference is again the presence of a large scepter extending from the head of the seahorse, along with hundreds of smaller scepters along the perimeter of the body.

> Each of the seahorses now has a prominent scepter extending from its head.

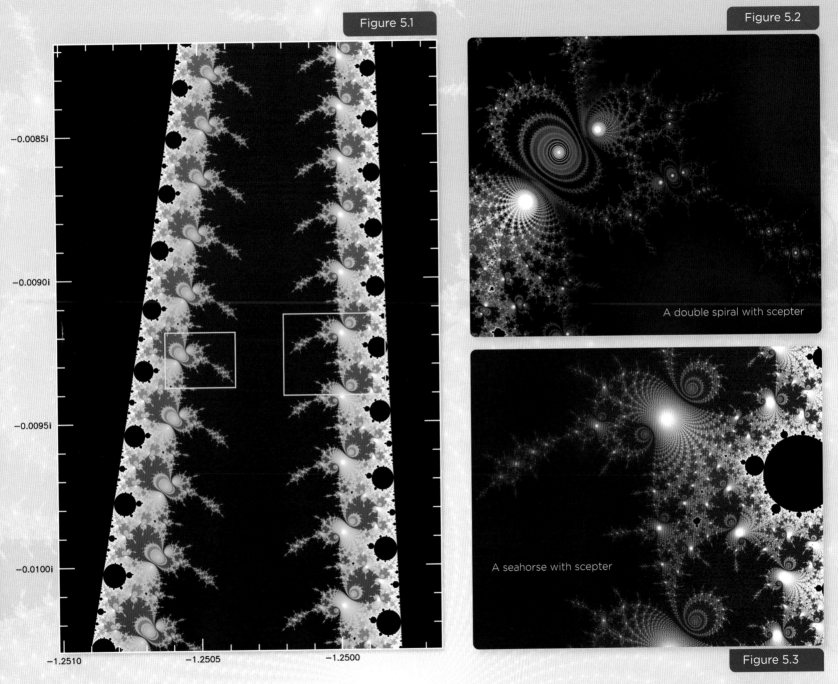

Figure 5.1

Figure 5.2

A double spiral with scepter

A seahorse with scepter

Figure 5.3

SCEPTORS ON SEAHORSES 65

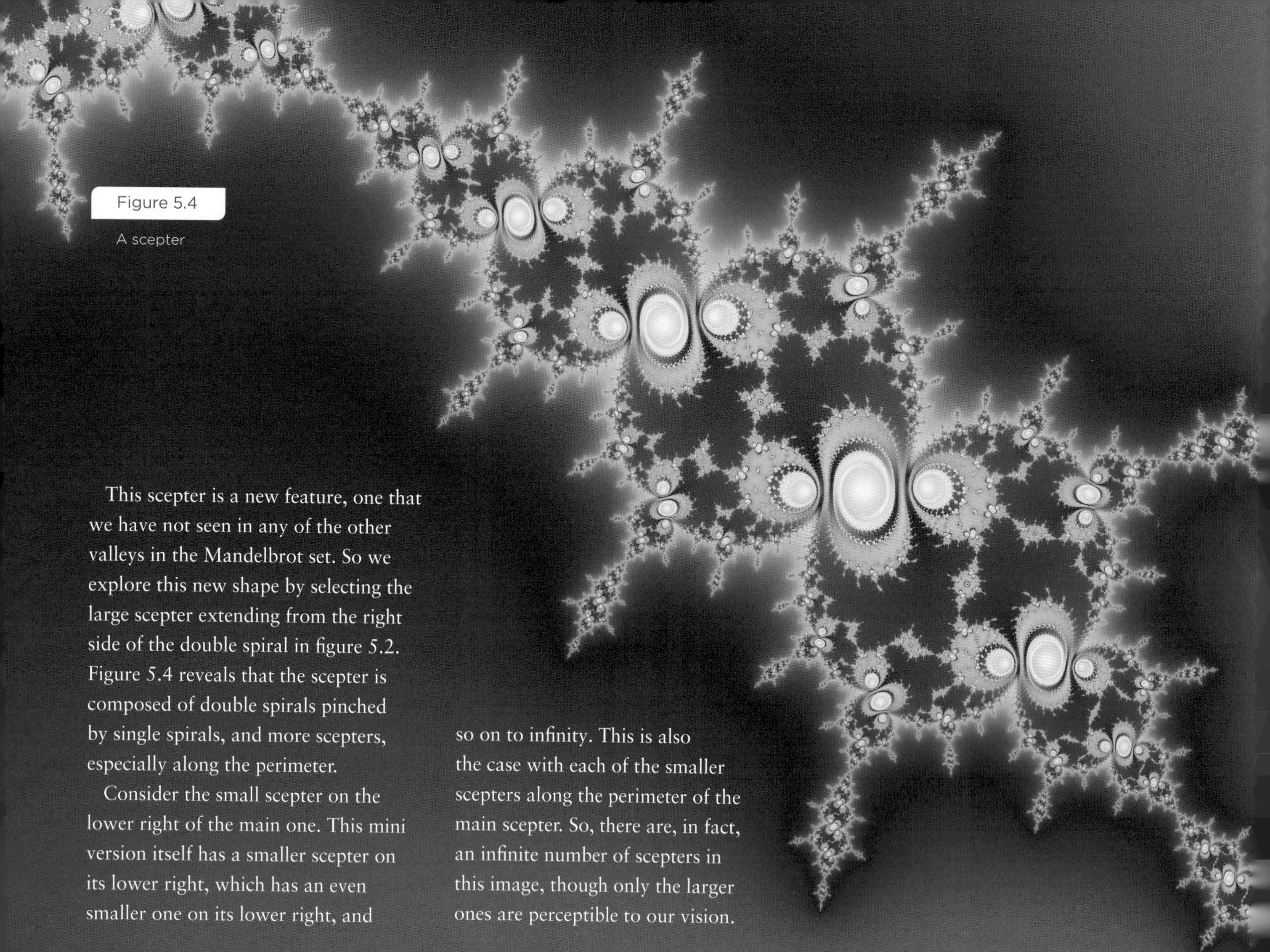

Figure 5.4

A scepter

This scepter is a new feature, one that we have not seen in any of the other valleys in the Mandelbrot set. So we explore this new shape by selecting the large scepter extending from the right side of the double spiral in figure 5.2. Figure 5.4 reveals that the scepter is composed of double spirals pinched by single spirals, and more scepters, especially along the perimeter.

Consider the small scepter on the lower right of the main one. This mini version itself has a smaller scepter on its lower right, which has an even smaller one on its lower right, and so on to infinity. This is also the case with each of the smaller scepters along the perimeter of the main scepter. So, there are, in fact, an infinite number of scepters in this image, though only the larger ones are perceptible to our vision.

Figure 5.5

Figure 5.5 is zoomed in on the center of this scepter, revealing two large double spirals on the upper left and lower right. Each of these is between two single spirals on the left and right. Note that the threads comprising these spirals have hundreds of scepters along their exteriors. In the exact center of this figure is a cross or "plus sign" structure that connects an upper spiral to a lower one. We zoom in on this cross to reveal figure 5.6.

Unsurprisingly, we find a miniature Mandelbrot in the center of this cross. We see the four largest double spirals at the top, bottom, left, and right of the plot. Closer to the center, we see 8 double spirals, then 16, and so on, just as we observed surrounding the miniature Mandelbrots in Seahorse, Double Spiral, and Elephant Valleys. The Mandelbrot set exhibits powers of two in its various shapes, apparently a result of the z^2 in the defining formula.

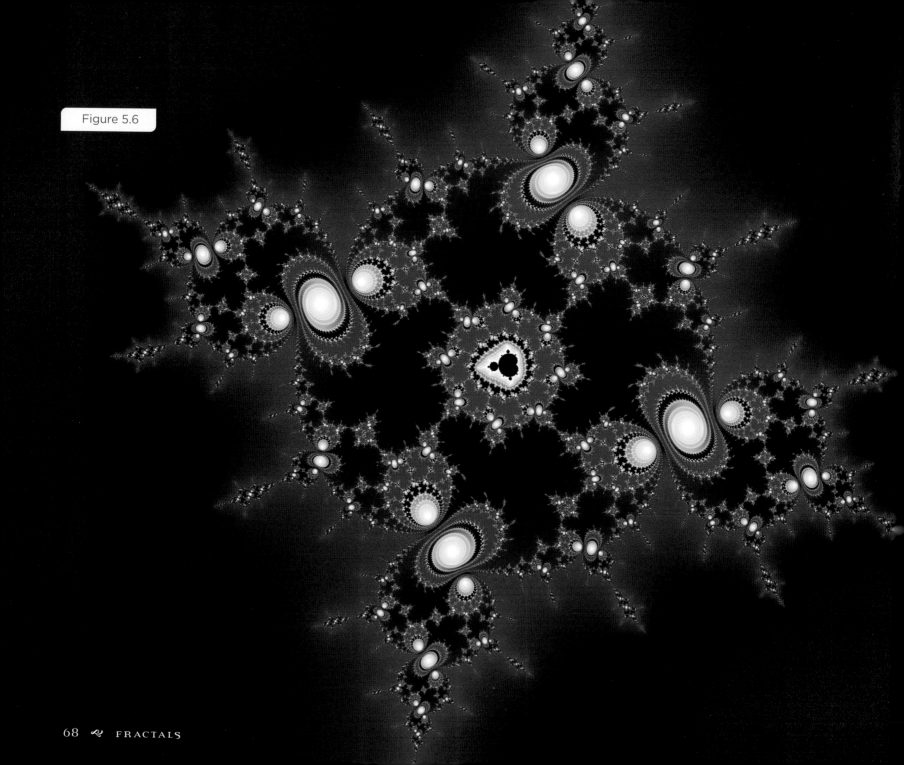

Figure 5.6

Implications

What are we to make of the astonishing beauty and complexity found in this map of the Mandelbrot set? How do we explain it? I have suggested previously that God is ultimately responsible for this shape. God's infinite mind determines mathematical truths, which necessarily include the truth of which numbers belong or do not belong to the Mandelbrot set. But, of course, not everyone agrees with that. What are the secular possibilities?

When the Mandelbrot set was first discovered, some scholars believed its self-repeating shape was an artifact of the computer code. In other words, the approximations made by the computer (such as finite numbers of iterations) were allegedly responsible for these artifacts, which would not exist if computers had the ability to iterate to infinity. But this view has been refuted. Mathematicians have demonstrated that the main regions of the Mandelbrot set (such as the main cardioid and the main disk) are truly bound, and the scale-invariance in the mini Mandelbrots is real. In other words, these amazing shapes really are built into numbers and are not simply a computer oddity.

Could human beings somehow be responsible for the patterns we have seen here? It seems absurd on the face of it. No human being could possibly create a structure of infinite complexity — we don't have the time. Indeed, no human decided that the Mandelbrot should consist of a cardioid with an infinite number of circles on its perimeter, which branch off into tendrils and so on. On the contrary, human beings were *surprised* by the shape that resulted when the Mandelbrot was finally plotted.

However, there is one aspect of the Mandelbrot set that was unquestionably determined by human beings: the formula. People decided to investigate the formula $z^2 + c$, and we have explored the map of those numbers that are bound under that formula. The map is completely determined by the rules of mathematics for the selected set. But we selected the set. Did we somehow create the above shapes by picking just the right formula?

This question prompts us to ask another: what happens if we change the formula? Is $z^2 + c$ somehow special? Did we serendipitously discover a unique formula in mathematics? If we plot a map of numbers that remain bound under a different formula, what will that look like? Will it be bland, or will it be infinitely rich in complexity like the Mandelbrot set, or will it be something in between? This will be the subject of the next chapter.

> God's infinite mind determines mathematical truths which necessarily include the truth of which numbers belong or do not belong to the Mandelbrot set.

God's infinite mind determines mathematical truths which necessarily include the truth of which numbers belong or do not belong to the Mandelbrot set.

Our investigation of the Mandelbrot set has led to some amazing discoveries. A simple plot of those points that are bound under the iteration $z^2 + c$ has revealed a structure of infinite complexity and astonishing beauty. The map contains double spirals that continue inward infinitely, along with seahorses, elephants, and scepters. Some of these shapes repeat on smaller scales to infinity, forming a fractal. But is this set unique? Are there other iterations whose map of bound points is fractal?

We begin by exploring what happens when we add a constant to the formula. Our new set is: $z^2 + c + 1$. A map of all numbers that are bound under this iteration is plotted in orange in figure 6.1. When compared to the standard Mandelbrot formula shown in black, we see the new set is identical in shape but shifted one unit in the negative x direction. Similarly, the set $z^2 + c - 1$ (magenta) is identical in shape but shifted one unit in the positive x direction. Adding an imaginary number shifts the shape in the y-direction, as shown in the blue and red plots. So the addition of any complex number to the formula merely shifts the map in the opposite direction. But the shape remains the same.

Figure 6.1

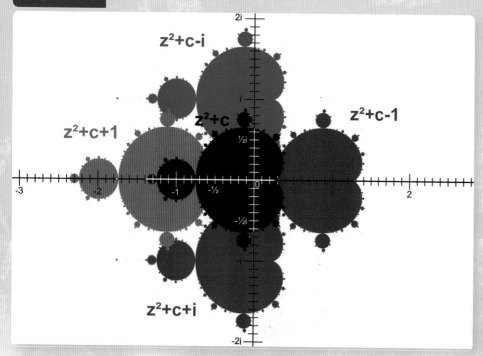

We next try variations in which we divide or multiply c by a positive constant. The map of points in the bound set $z^2 + 2c$ is shown in red in figure 6.2. The shape is identical to the standard Mandelbrot but is exactly half the width and height. On the other hand, when we divide c by a positive integer, the resulting set is larger than the original by the amount of the dividend. A plot of numbers bound by $z^2 + c/2$ is shown in blue and is exactly twice the height and width of the original.

What happens when we multiply c by negative or imaginary numbers? The map of bound points under the iteration $z^2 - c$ is shown in magenta in figure 6.3. It is a mirror image of the Mandelbrot — reversed in the x direction. Multiplying c by an imaginary number rotates the map by 90 degrees. The iteration $z^2 + ic$ maps to a Mandelbrot set rotated by 90° in the clockwise direction (red), whereas the iteration $z^2 - ic$ rotates the map 90° counterclockwise (blue). Multiplying c by a complex number rotates the map by an intermediate angle.

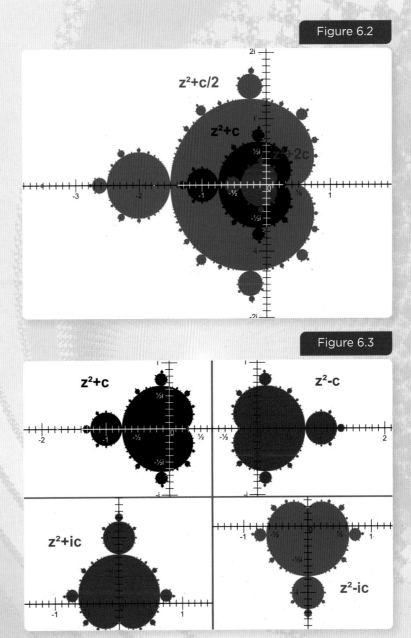

Figure 6.2

Figure 6.3

CHANGING THE FORMULA 73

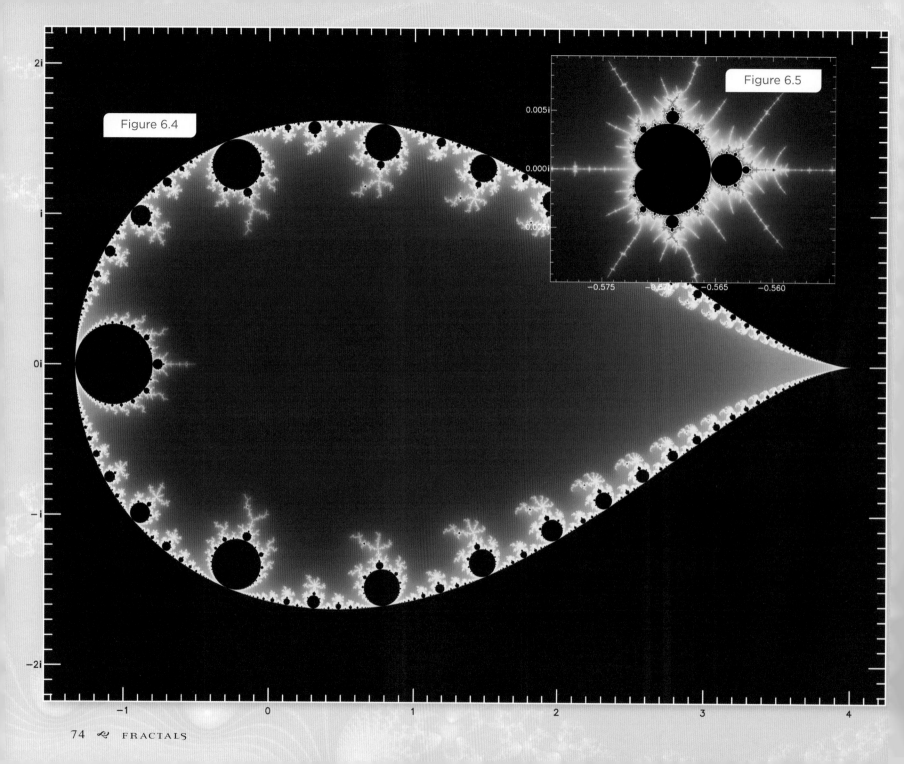

Figure 6.4

Figure 6.5

Apparently, any additive or multiplicative change to c has no effect on the shape of the resulting map. The results are simply a shifted, scaled, or rotated Mandelbrot map. The shape of all internal structures is unaffected. Each of these variations has a Mini on the main spike, a Seahorse Valley, a Double-Spiral Valley, an Elephant Valley, a Scepter Valley, and so on.

A more interesting situation occurs when we take c to some power. Consider c to the power of negative one (c^{-1}), which is the reciprocal of c. The map of this iteration ($z^2 + 1/c$) is shown in figure 6.4. This change essentially flips the Mandelbrot map inside-out. Notice that the points that belong to this inverted Mandelbrot set (shown in black) are now on the *exterior*, whereas those points that do not belong form the interior of the raindrop. So this inverted Mandelbrot has infinite area, whereas the area of the points that do *not* belong is finite — the opposite of the standard Mandelbrot map.

The main cardioid, which had one cusp on the right side, has now become an inverted raindrop, also with one cusp. And instead of an infinite number of circles attached to the *exterior* of the cardioid, we now have an infinite number of circles *within* the perimeter of the raindrop. The shape truly is inverted, just as we would expect since we are considering the reciprocal of c.

Even the location of the substructures makes sense. The location of the end of the cusp for the Mandelbrot set was c = ¼. For the inverted Mandelbrot, the cusp ends at c = 4, which is, of course, the reciprocal of ¼. Likewise, the end of the spike on the left of the Mandelbrot set occurs at c = -2, whereas the spike on the inverted Mandelbrot (now extending to the *right* side of the main disk) ends at c = -½, which is the reciprocal of -2.

The (normal) Mandelbrot map had a mini-Mandelbrot near the end of the spike (figures 2.1, 1.4). When we zoom in on the equivalent spike of this inverted Mandelbrot map, we find that it too has a mini (figure 6.5). Interestingly, this mini is that of a *normal* Mandelbrot and is *not* an inside-out teardrop, although it is reversed left-to-right. Likewise, we find Elephant Valley in the cusp of the raindrop just as we found in the cusp of the cardioid, though the elephants are marching in the opposite direction. They are enlarged in this inverted Mandelbrot relative to their reversed cousins. Likewise, the cusp between the main disk and raindrop has a Seahorse Valley and Double Spiral Valley — reversed left-right from their counterparts. One of the seahorses is plotted in figure 6.6. Aside from being reversed, it appears essentially identical to its counterpart in the normal, non-inverted Mandelbrot.

Apparently, the internal structures of this new shape are the same as those in the original Mandelbrot set. Although

> Apparently, the internal structures of this new shape are the same as those in the original Mandelbrot set.

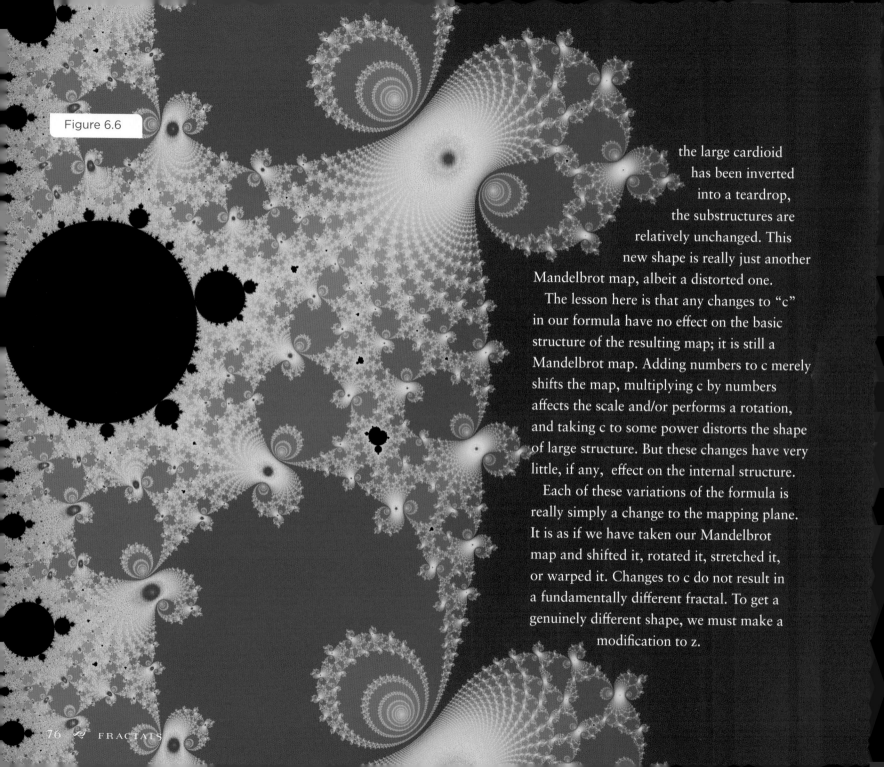

Figure 6.6

the large cardioid has been inverted into a teardrop, the substructures are relatively unchanged. This new shape is really just another Mandelbrot map, albeit a distorted one.

The lesson here is that any changes to "c" in our formula have no effect on the basic structure of the resulting map; it is still a Mandelbrot map. Adding numbers to c merely shifts the map, multiplying c by numbers affects the scale and/or performs a rotation, and taking c to some power distorts the shape of large structure. But these changes have very little, if any, effect on the internal structure.

Each of these variations of the formula is really simply a change to the mapping plane. It is as if we have taken our Mandelbrot map and shifted it, rotated it, stretched it, or warped it. Changes to c do not result in a fundamentally different fractal. To get a genuinely different shape, we must make a modification to z.

The Multibrot

Since the Mandelbrot set is defined by the formula $z^2 + c$, we found several effects within the structure that were related to the number 2. Namely, the number of similar structures surrounding the tiny mini-Mandelbrots were always a power of 2, as shown in figure 5.6. In that figure, there are 4 double spirals around the perimeter, with 8 deeper in, then 16, and so on. So, what happens when we change the power of z?

Let's consider what happens when we change the power of z from 2 to 3. A map of those points bound under the iteration $z^3 + c$ is shown in figure 6.7. Although there are some similarities to the map of the Mandelbrot set, this is a fundamentally different shape. The Mandelbrot set is symmetric about the x-axis, but is not symmetric about the y. However, this new shape is symmetric both along the x and y axes. And the substructures are also changed, as we will see.

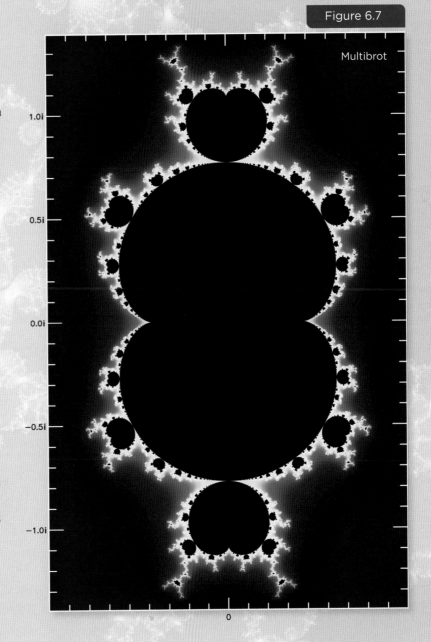

Figure 6.7

Multibrot

CHANGING THE FORMULA 77

So the multibrot set, like the Mandelbrot set, is a fractal. It contains miniature versions of itself on increasingly smaller scales.

The differences between this new shape and the Mandelbrot map make mathematical sense. We have increased the power of z by one, and many of the substructures in this new shape have increased by one. For example, the main cardioid in the Mandelbrot set had exactly one cusp on the right side. The largest structure on this new set has exactly two cusps: one on the right (as before) and a new one on the left. This new double-cusped shape is called a *nephroid*. Just as a cardioid is what you get when you roll one circle around another of equal size, a nephroid results when you roll a circle around another twice its size.

Other structures have also increased by one. The circles (which have zero cusps) growing off the cardioid of the Mandelbrot map have become cardioids (which have one cusp each) growing off the nephroid of this new map. In a sense, this new shape is a *multiple* of the Mandelbrot set and is referred to as a *multibrot*.

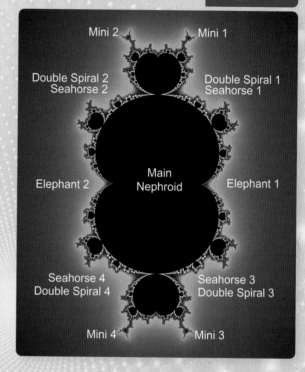

Figure 6.8

The main disk on the Mandelbrot has also doubled in this new map. It has split and morphed into two cardioids at the top and bottom of the nephroid. (We will examine how we can track the morphing of one set into another in the next chapter.) But there are some exceptions to this doubling principle. The "mini" sprouting from the spike on the

Mandelbrot set has split into *four* copies on this new shape (see figure 6.8). But when we zoom in on them, will they be mini-Mandelbrots, or mini-*multibrots*?

Figure 6.9 is a plot of the upper right-hand mini (Mini 1), and it is indeed a smaller version of the multibrot. Several smaller mini-multibrots are also visible. So the multibrot set, like the Mandelbrot set, is a fractal. It contains miniature versions of itself on increasingly smaller scales. As with the Mandelbrot mini, extra branches extend from the perimeter of this mini, which were not present on the larger structure. The miniature versions gain additional complexity and inherit the properties of the section of the main set on which they are found.

What has become of the various valleys we explored on the Mandelbrot set? The structure that seems least changed in this new set is the cusp on the right-hand side of the main nephroid, which we label "Elephant 1" in figure 6.8. Its position and shape are very similar to the cusp in the main cardioid of the Mandelbrot set, where we found Elephant Valley. Will we find elephants in the multibrot set as well?

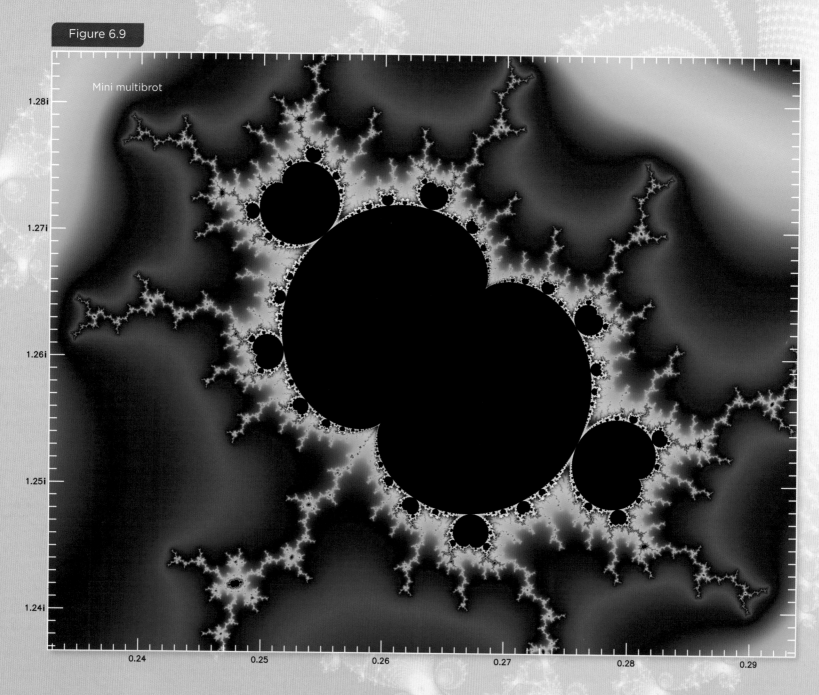

Figure 6.9

Mini multibrot

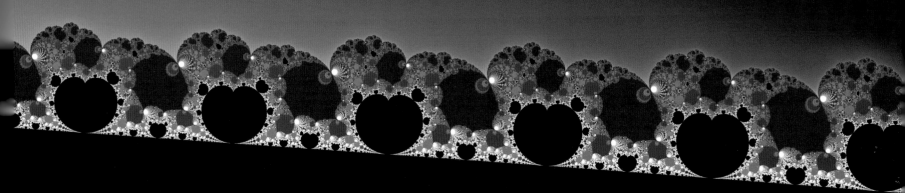

Figure 6.10

Elephant Valley of the multibrot

Multibrot Elephants

In figure 6.10, we have zoomed in deep in the cusp of the nephroid, and indeed we find elephants. But there is a difference. Recall that Elephant Valley in the Mandelbrot set had one elephant standing on each circle (figure 4.2). Here in the multibrot, each circle has become a cardioid, and there are now *two* elephants on each cardioid. The elephants have doubled. This is amazing when we consider that there were an infinite number of elephants in the Mandelbrot set, and now we have twice that many! Infinity is such a wonderful and counterintuitive concept. But this is the way God thinks, and it should produce in us a profound sense of humility.

Not only have the elephants in Elephant Valley doubled, but the Valley itself has doubled. The multibrot has two Elephant Valleys, one on the right and a new one on the left (figure 6.8). As you might suspect, Elephant Valley 2 is a mirror image of Elephant Valley 1.

Figure 6.11 is a plot of the lower central elephants of the previous figure. The elephant on the left of the cardioid appears a bit larger but with a smaller trunk than his brother on the right. This is a common property of multibrot elephants. But how does the internal structure of these elephants compare to those of the Mandelbrot set?

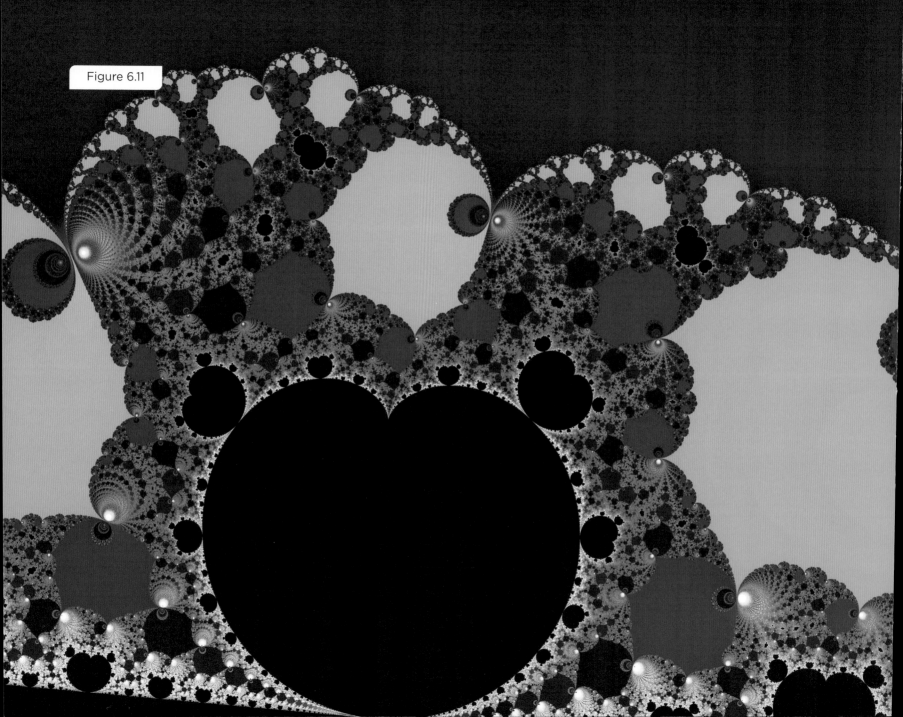

Figure 6.11

Figure 6.12

An enlarged plot of the trunk of the left elephant is plotted in figure 6.12. It is a single spiral and very similar to the coiled trunk of the elephants in the Mandelbrot set (compare with figures 4.4 and 4.5). Figure 6.13 is a zoom-in on one of the strands of this spiral. We see more single spirals and spiderweb structures, along with some new 3-fold structures. These new structures are similar to the "bowties" in the Mandelbrot set, but they now have a three-fold rather than a two-fold structure.

CHANGING THE FORMULA 83

Figure 6.13

Zooming in on the exact center of figure 6.13 reveals figure 6.14, which shows one of these "tri-ties" in all its beauty. Comparison with figure 4.8 shows that this tie has gained an additional spiral from its Mandelbrot counterpart. The three single spirals surround nine white smaller spirals near the core. Recall that the Mandelbrot spirals go from 2, to 4, to 8, to 16, and so on — all the powers of 2. This new tie has spirals which go from 3, to

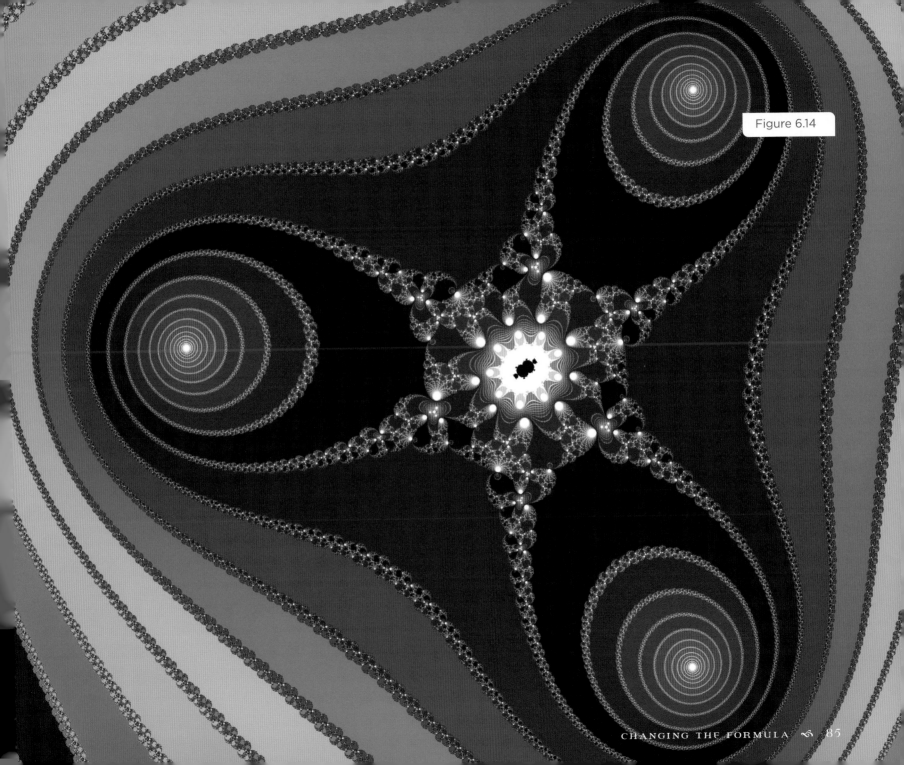
Figure 6.14

CHANGING THE FORMULA 85

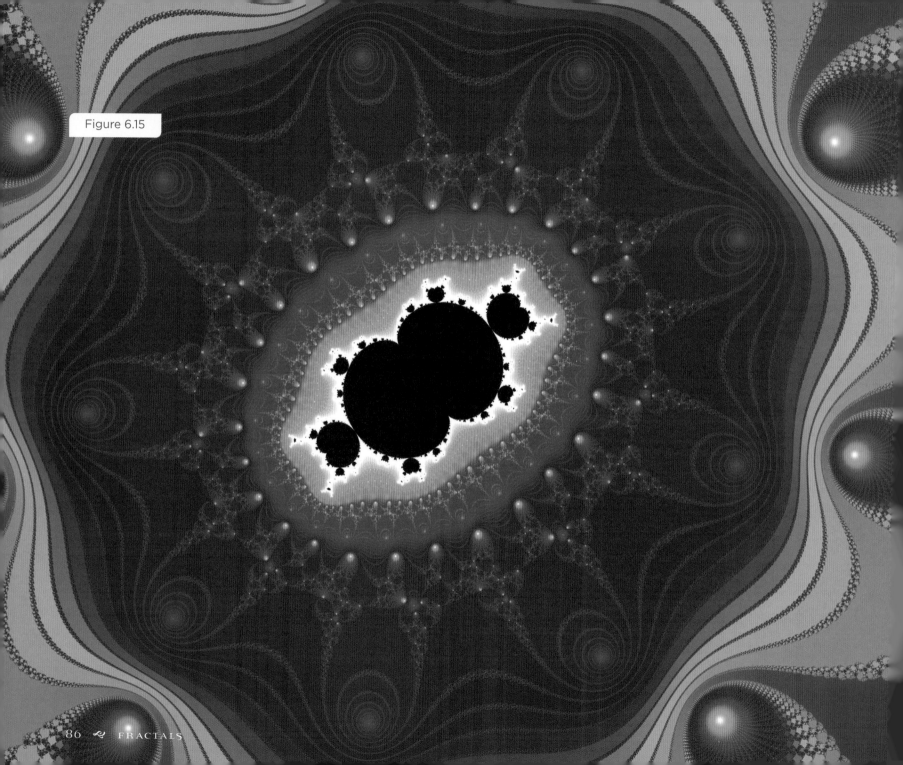

Figure 6.15

9 (figure 6.14), to 27, to 81 (figure 6.15), and so on — all the powers of 3. This, of course, is due to the fact that z is taken to the third power in this multibrot. In the center is a miniature version of the multibrot, shown in detail in figure 6.15.

This tiny multibrot resembles the basic structure of the larger version (compare with figure 6.7), but it has a far more intricate exterior. Single spirals surround this mini, which itself was found in a single spiral. This shows that baby versions of multibrots inherit the characteristics of the section of the parent on which they are found.

Exploring the Valleys

The other valleys in the Mandelbrot set also have their counterparts in the multibrot, although their locations have shifted (see figure 6.8). Moreover, the valleys themselves have multiplied. The Seahorse and Double Spiral Valleys have each gone from two to four. Figure 6.16 shows the interior of Seahorse Valley (1) on the bottom, and Double-Spiral Valley (1) on the top.

As with Elephant Valley, the circles in the Mandelbrot version of Seahorse Valley and Double Spiral Valley have become cardioids in the multibrot. Each seahorse and double spiral has doubled, with two on each cardioid. The left-hand seahorse on each cardioid is larger than the right-hand one, but the right-hand one has a larger tail (see figure 6.17). These kinds of patterns are common in the multibrot.

The double spirals are similarly duplicated in the multibrot set. Zooming in on one of them (figure 6.18) shows that they strongly resemble their counterparts in the Mandelbrot set (compare with figure 3.4), and mini versions of the entire set are found within. But of course, these minis resemble the multibrot with its main nephroid. So, although the multibrot is a fundamentally different fractal from the Mandelbrot, they share many common features, with the multibrot apparently gaining copies of these small-scale structures.

Figure 6.16

Figure 6.17

Figure 6.18

What happens when we try other powers of z? In general, how is the overall shape of the fractal affected by the power of z? And is there any worldview besides the biblical one that can make sense of these intricate patterns found in math?

CHANGING THE FORMULA

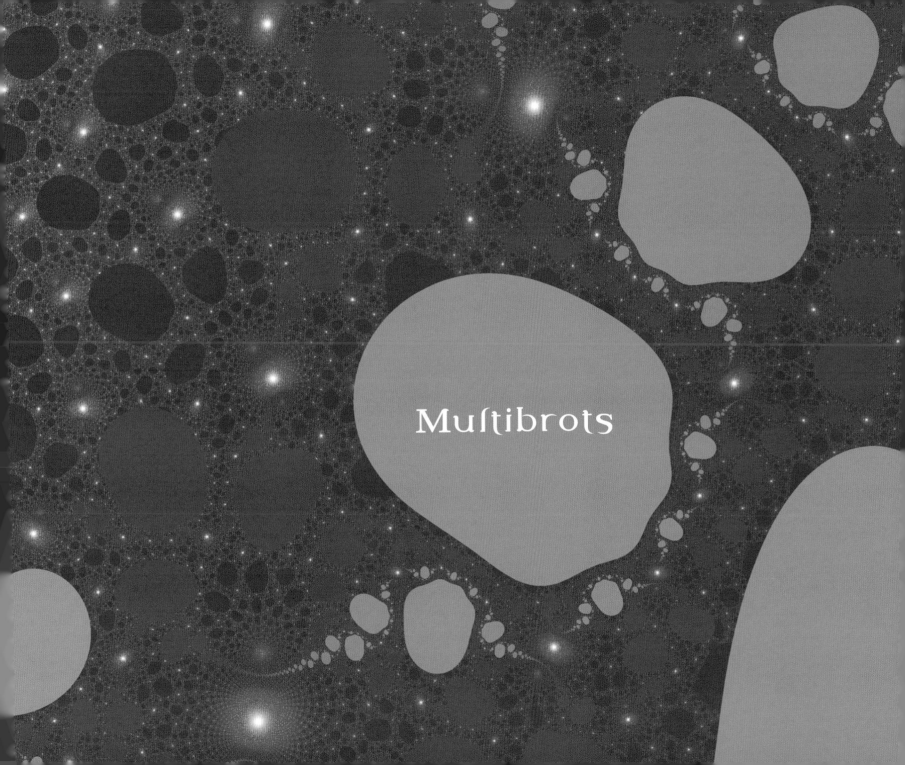

We have seen that changing the power of z in the Mandelbrot formula results in an entirely new map: a multibrot. Just as we could spend eternity investigating the infinite depths of the Mandelbrot set, we could likewise explore the multibrot forever without reaching the inexhaustible depths of its beauty. And yet, there are other powers of z we might explore. There are, in fact, an infinite number of them and therefore an infinite number of multibrots — each of which is infinitely complex. It would seem that the Mind responsible for fractals is not only infinite, but *infinitely* infinite!

Each multibrot is an infinite universe for us to explore. We might call the version we explored in the previous chapter the *first* multibrot since its formula is the next power of z. Our goal in this chapter will be to explore other multibrots and see how the power of z affects the overall map of the set as well as its substructures.

> Each multibrot is an infinite universe for us to explore.

Having explored $z^2 + c$ and $z^3 + c$, we will continue the progression and explore the set of all points bound under the formula $z^4 + c$. The map of this multibrot is shown in figure 7.1. Just as the first multibrot in the last chapter had a two-fold symmetry, this second multibrot has a three-fold symmetry.

As before, incrementing the power of z by one increases many of the substructures by one. The large central section now has three cusps, just as the Mandelbrot had one and the first multibrot had two. But some structures are duplicated even more. The mini Mandelbrot that increased to four on the first multibrot has increased to nine in this second multibrot.

Zooming in on the far left mini multibrot, we see that this second multibrot is also fractal (figure 7.2). The mini resembles the parent in every way, except for extra dendrites that sprout along the perimeter. These dendrites contain several even smaller versions of this multibrot.

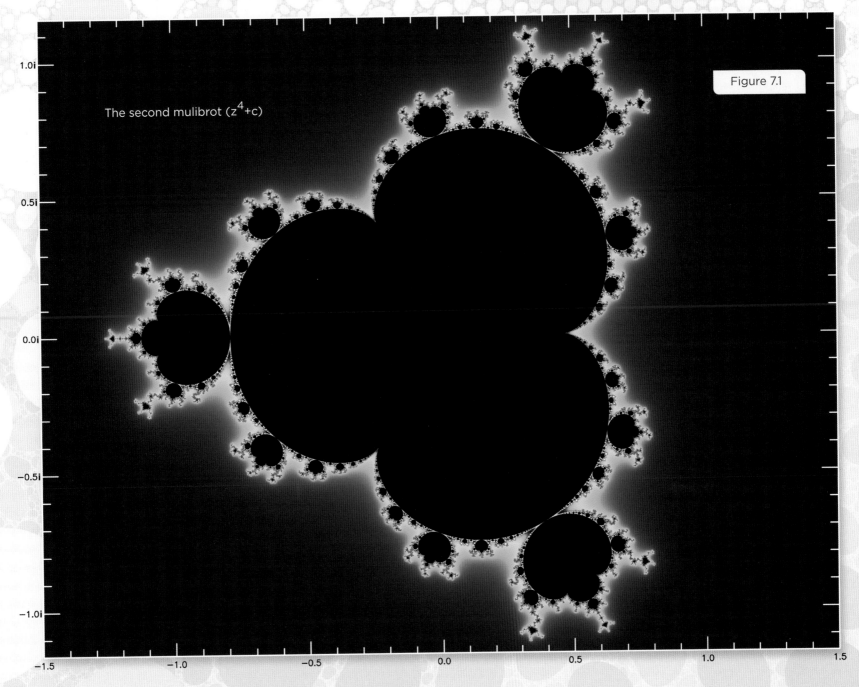

Figure 7.1 — The second mulibrot (z^4+c)

MULTIBROTS 93

Figure 7.2

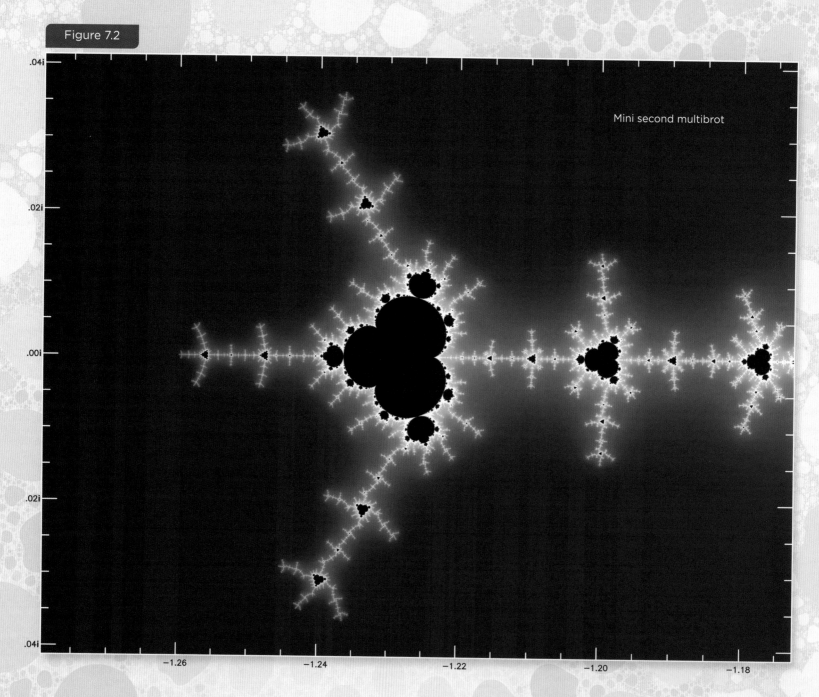

Mini second multibrot

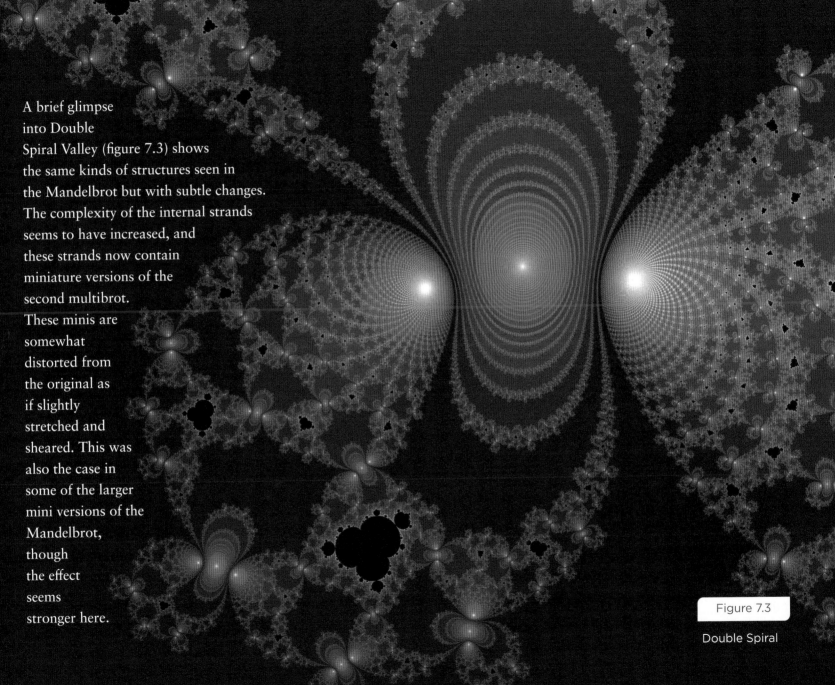

A brief glimpse into Double Spiral Valley (figure 7.3) shows the same kinds of structures seen in the Mandelbrot but with subtle changes. The complexity of the internal strands seems to have increased, and these strands now contain miniature versions of the second multibrot. These minis are somewhat distorted from the original as if slightly stretched and sheared. This was also the case in some of the larger mini versions of the Mandelbrot, though the effect seems stronger here.

Figure 7.3

Double Spiral

Figure 7.4

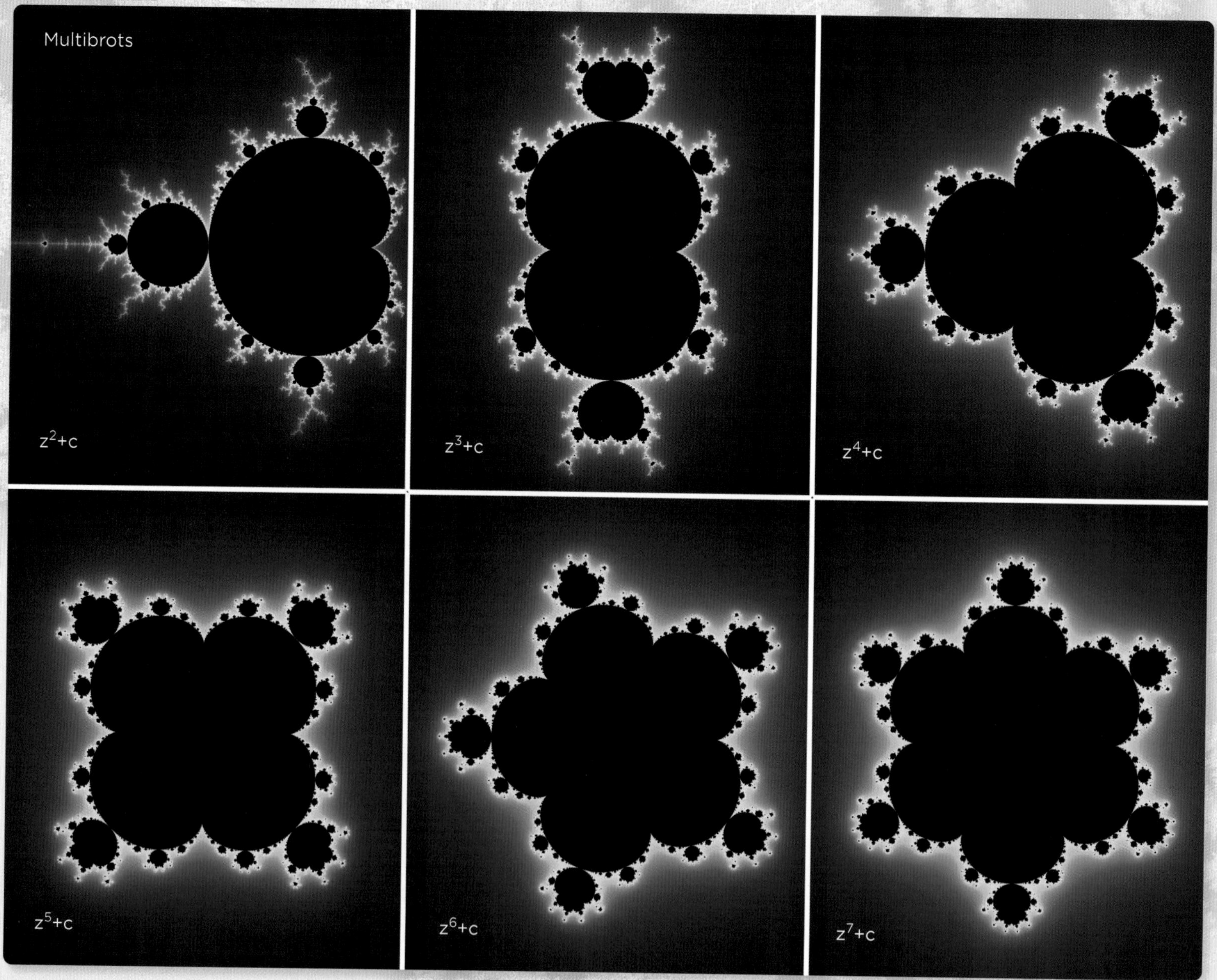

Multibrots

z^2+c

z^3+c

z^4+c

z^5+c

z^6+c

z^7+c

Apparently, multibrots of any positive finite power contain the same kinds of structures seen in the Mandelbrot. But these features become duplicated and with increasing levels of complexity as the power of z is incremented. So let us consider how the overall map of the set is affected by comparing multibrots of increasingly high powers.

The first three panels of figure 7.4 show the entire map of the Mandelbrot and the first and second multibrots respectively for easy comparison. The next three panels show the next three multibrots, $z^5 + c$, $z^6 + c$, and $z^7 + c$. The pattern becomes clear. Each multibrot has one more cusp in its central shape than the previous. Each has a symmetry that is one greater than the previous and one less than the power of z. Hence, the map of $z^4 + c$ has a three-fold symmetry, and the map of $z^5 + c$ has a four-fold symmetry. Since $z^7 + c$ has sixfold symmetry, it is reminiscent of the structure of a snowflake. Of course, mini versions of this (fifth) multibrot are also snowflakes (see figure 7.5).

Figure 7.5

NON-FRACTAL SETS

For the sake of curiosity, we also plot the map of $z^\infty + c$ (figure 7.6). That is z to the power of infinity. Curiously, this map is *not* a fractal but is instead a perfect circle. This is wonderfully ironic: the *finite* powers of z yield maps of *infinite* complexity, whereas the *infinite* power of z yields a map that is *finite* and simple.

Figure 7.6

This map of $z^\infty + c$ presents a circle with a radius of one, centered exactly on zero. This makes sense because when any number greater than one is taken to the power of infinity, the result is infinity. Conversely, any number less than 1 when taken to the power of infinity is zero. So in one iteration, any point exterior to the unit circle will have a z value that escapes to infinity.

Other non-fractal sets include those in which z is taken to the power of one or zero. The map of $z^1 + c$ (which is simply $z + c$) results in a set that is empty except for one point. The number zero is the only member of this set because any other number results in a sequence of z that is unbound.

Conversely, the map of $z^0 + c$ is the set of all numbers. Every value of c results in a sequence z_n in which all values (except n = 0) are the same and are therefore bound. However, if we plot this set using an escape limit as we did for the Mandelbrot, then the resulting map will be a perfect circle, centered on c = -1 with a radius equal to the escape value. This is because $z^0 = 1$ for any value of z.

Figure 7.7

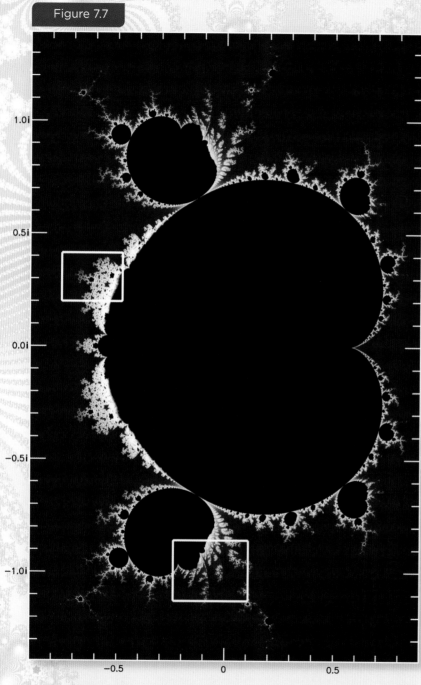

"Broken" Fractals

We might also consider sets in which the power of z is not an integer. For example, consider the map of the set of all points bound under the formula $z^{2.5} + c$. The power of z is halfway between that of the Mandelbrot and the first multibrot. Unsurprisingly, the shape of the resulting map is qualitatively in between the shapes of those other two maps (figure 7.7). The central shape is not quite a cardioid, but neither is it a nephroid. The second largest structures are no longer circles, but they are not quite cardioids either. Many of the substructures of this unhappy fractal appear to be broken, in contrast to the perfect geometric shapes seen in the integer-powered multibrots.

We enlarge the region of the upper box and find a frostlike structure with several miniature multibrots (figure 7.8). But notice that these miniature versions do not match the shape of the original. Rather, they each appear to be distorted combinations of the Mandelbrot and the first multibrot. However,

Figure 7.9

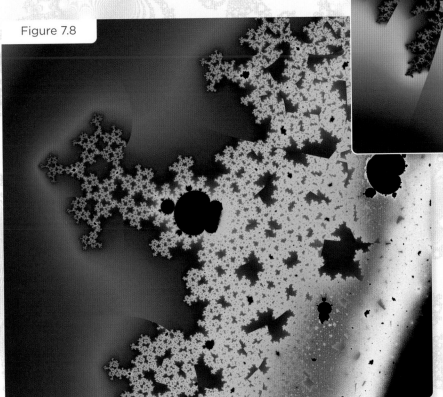

Figure 7.8

Figure 7.10

they are distorted in different ways from the overall map and from each other. Each mini looks unique and yet captures some features of the Mandelbrot and the first multibrot. It is as if this tortured fractal would really prefer to be either the Mandelbrot or the first multibrot, but its substructures only result in a mutated amalgamation of both. Apparently, only the integer-powered multibrots contain smaller versions that match the shape of the entire map.

We can see the "broken" nature of this fractal by enlarging the lower boxed region of figure 7.7. Figure 7.9 reveals a shattered version of the kinds of patterns found in the multibrots. But we do see a new unique feature: "islands." Recall that the Mandelbrot set is connected. We cannot draw a loop around any partial region of the Mandelbrot without intersecting the Mandelbrot. This is also true of the integer-powered multibrots. But fractional-powered multibrots can have numerous islands that are spatially isolated from the main body.

The shattering pattern along with the isolated miniatures open up some new and beautiful structures for our exploration and appreciation (figure 7.10), unseen in the Mandelbrot. So we again have a new and unique universe of infinite depth to explore. And, of course, there are an infinite number of other fractional-powered multibrots, each of infinite complexity. But our goal here is to get an overview of all multibrots. So we'll move on to another category of multibrots whose maps are of an entirely different nature.

> Apparently, only the integer-powered multibrots contain smaller versions that match the shape of the entire map.

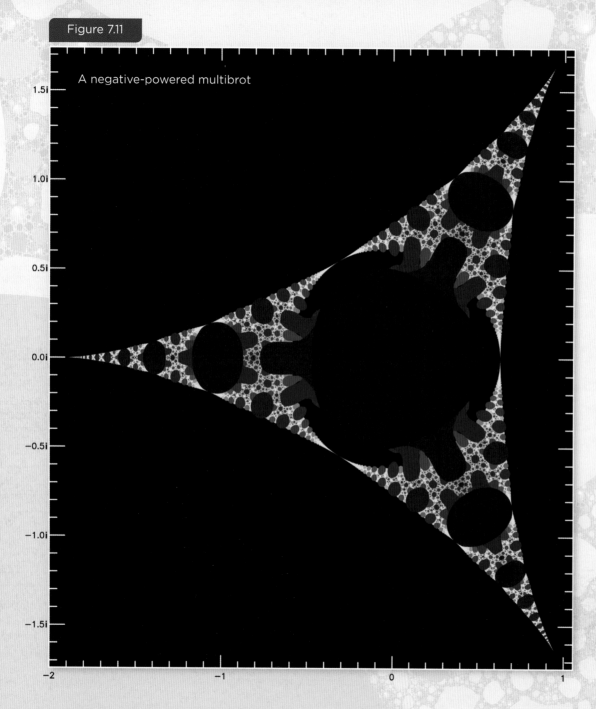

Figure 7.11

A negative-powered multibrot

Negative Multibrots

What happens when we consider multibrots that have a negative power of z? A negative power such as z^{-2} is simply the reciprocal of z^2. That is, $z^{-2} = 1/z^2$. In order to examine these sets, we need to make at least one small adjustment to our mapping algorithm. Recall, we are going to examine the sequence z_n under the following iteration:

$$z_n^{-2} + x = z_{n+1}$$

For the Mandelbrot set, we said that the first value of z is zero by definition: $z_0 = 0$. But this will not work for negative multibrots because zero to any negative power forces us to divide by zero, which instantly escapes to infinity. So let us set the first value of z to c, rather than zero. (In fact, we could have used this alternate definition for the Mandelbrot and the positive-powered mutlibrots as well without affecting the resulting maps. This is because z_1 is always equal to c anyway if z_0 = 0 for positive powers.)

There is another mathematical caveat we will address later concerning the bailout criterion. For now, we simply set the bailout number to 4, and we set z_0 = c. We can then explore the map of z^{-2} + c by the escape time algorithm just as with the other multibrots.

The map of z^{-2} + c is shown in figure 7.11. We are delighted to find an entirely new structure, completely dissimilar to any of the positive-powered multibrots. We first note that the map is *inverted* with respect to its positive-powered counterparts. The black area — representing points that do belong to the set — is now mostly exterior to the central region. That is, nearly the entire infinite Argand plane belongs to this inverted Mandelbrot set. It has infinite area. However, the points that do not belong to the set are now on the *inside* of the perimeter and have finite area.

A second notable difference is that while the Mandelbrot set had an infinitely wiggly perimeter, the perimeter of this inverted set appears relatively smooth and non-fractal. In fact, it approximates the intersection of three circles.

Finally, the interior region — representing points that do not belong to the set — appears as a pattern of overlapping pebbles. We have enlarged one section of this negative mulibrot in figure 7.12. This delightful cobblestone pattern is not seen in the positive multibrots, but we will find that it is a common feature of negative multibrots.

The size of these pebbles, however, is not an intrinsic property of the set. Rather, it is determined by the bailout criterion — larger bailouts result in smaller pebbles. This is quite different from positive multibrots where the internal structure is not affected by the bailout criterion, providing the number of iterations is sufficiently high. We'll explore this caveat in more detail later. The location of the pebbles, however, is not affected by the bailout criterion, and so these pebbles are an aspect of the set. Furthermore, the basic shape of the map is relatively unchanged by bailout criterion; the threefold symmetry remains.

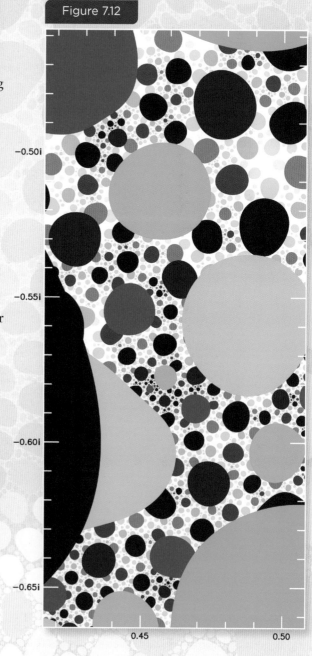

Figure 7.12

Figure 7.13

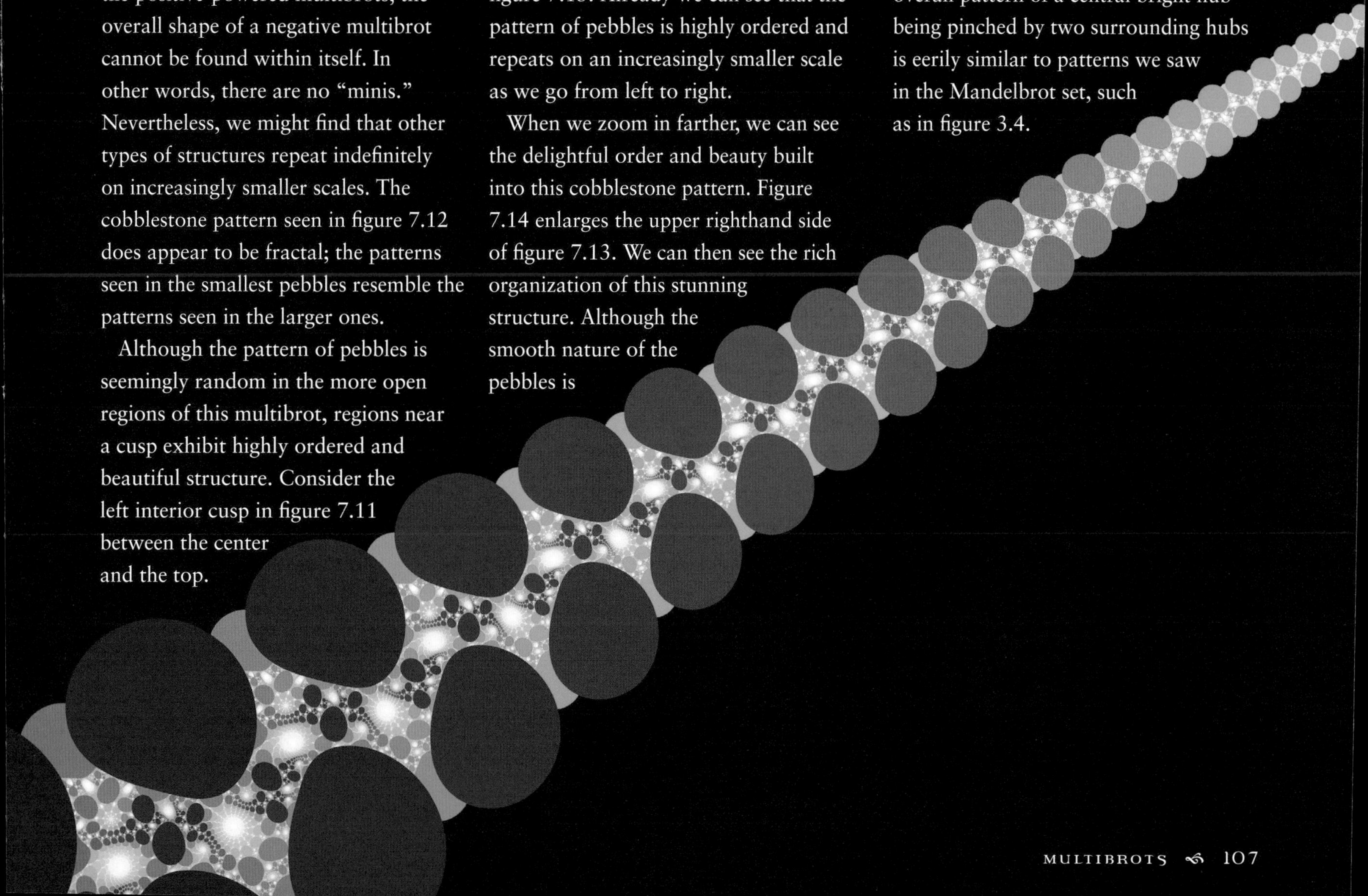

Are negative multibrots fractal? Unlike the positive-powered multibrots, the overall shape of a negative multibrot cannot be found within itself. In other words, there are no "minis." Nevertheless, we might find that other types of structures repeat indefinitely on increasingly smaller scales. The cobblestone pattern seen in figure 7.12 does appear to be fractal; the patterns seen in the smallest pebbles resemble the patterns seen in the larger ones.

Although the pattern of pebbles is seemingly random in the more open regions of this multibrot, regions near a cusp exhibit highly ordered and beautiful structure. Consider the left interior cusp in figure 7.11 between the center and the top.

This structure is shown enlarged in figure 7.13. Already we can see that the pattern of pebbles is highly ordered and repeats on an increasingly smaller scale as we go from left to right.

When we zoom in farther, we can see the delightful order and beauty built into this cobblestone pattern. Figure 7.14 enlarges the upper righthand side of figure 7.13. We can then see the rich organization of this stunning structure. Although the smooth nature of the pebbles is unique to negative multibrots, the overall pattern of a central bright hub being pinched by two surrounding hubs is eerily similar to patterns we saw in the Mandelbrot set, such as in figure 3.4.

MULTIBROTS 107

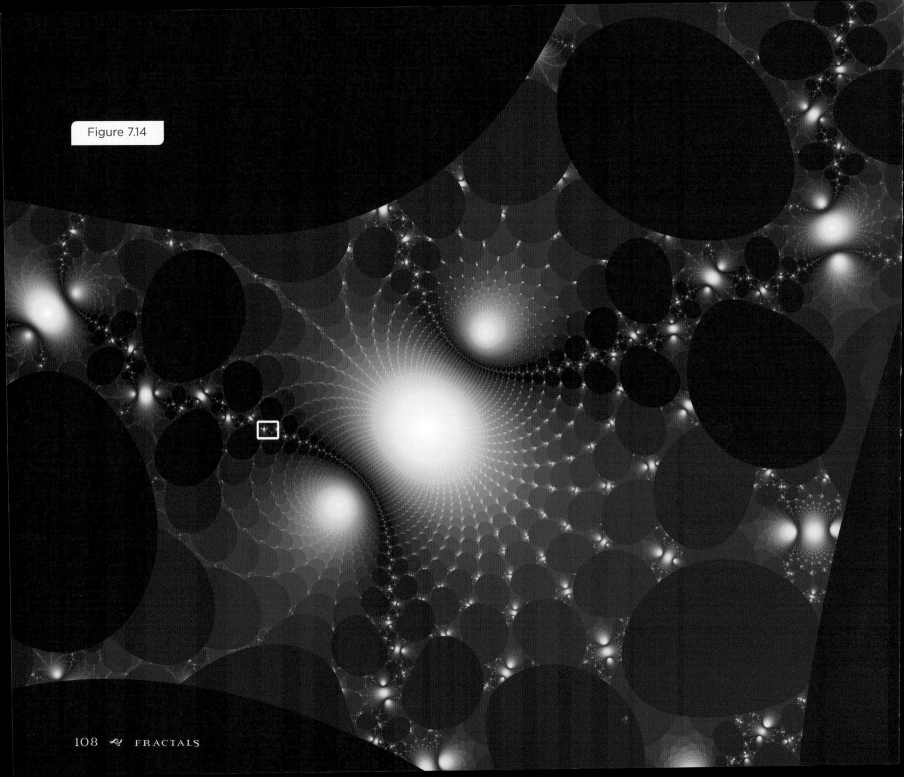

Figure 7.14

Note also that we are looking at a largely empty region of this set. There are no black points to be seen — no points that actually belong to this set are large enough to be detected. Individual points belonging to the set may well lie in this region, but they are isolated and have a total area of zero.

We also note that the fractal nature of this shape is increasingly clear. The pebbles surrounding the bright central hub decrease with size as we approach the hub's center and continue inward apparently to infinity.

The fractal nature of this shape becomes even more apparent when we enlarge the boxed region, as shown in figure 7.15. The structure is nearly unchanged, but on a much smaller scale. Clearly, the negative multribrot has scale-invariance. It is a fractal.

The way in which the pebbles appear to overlap gives negative multibrots a pseudo three-dimensional structure. Nearby pebbles appear to partially eclipse those in the background. However, this is merely an illusion. The shape is truly two-dimensional, having height and width but no depth. This is necessarily the case for all multibrots because they are plotted in the Argand plane, which is inherently two-dimensional.

The depth illusion is generated by our natural proclivity to assume that foreground objects block background objects and never the reverse. This is enhanced by the fact that the pebbles intersect, with the larger ones apparently cutting into the smaller ones. Our everyday experiences suggest that larger means closer, which reinforces the illusion. But, of course, good artists know how to create the appearance of depth on a two-dimensional canvas, so we shouldn't be surprised that the Great Artist also knows how to do this.

Before exploring other negative multibrots, we will briefly investigate the question of whether the above graphs represent the true appearance of these sets. There is no

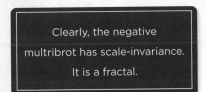

Clearly, the negative multribrot has scale-invariance. It is a fractal.

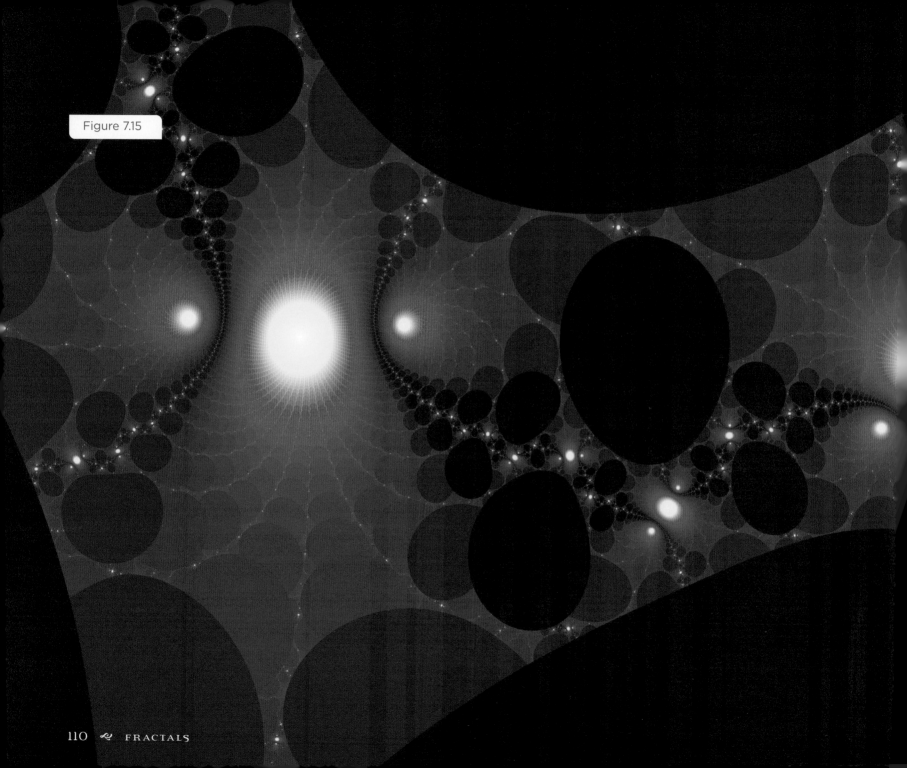
Figure 7.15

doubt that when the computer follows the above algorithm, these shapes are what emerge. But some people have questioned the appropriateness of using the escape time algorithm with negative-powered sets. And this deserves a brief discussion.

Recall that the Mandelbrot and multibrot sets represent those sets of points c where the iteration of z is bound under the given formula ($z^p + c$, where p depends on which multibrot we are plotting). And z being "bound" means that the sequence does not "run away" — that even though there are an infinite number of z_n, they never exceed a certain maximum value. For the Mandelbrot set, it has been mathematically proven that if any value z_n is greater than 2, then the sequence is unbound — it increases indefinitely, and hence, there is no maximum value of z_n. So the number 2 is the "escape radius" or "bailout criterion." You don't need to continue the algorithm after z_n exceeds the escape radius because you know the sequence is unbound.

This method of plotting a fractal is called the "escape time algorithm" or "bailout method." And it makes sense for positive-powered multibrots because at large values of z, z^p will be even larger than z and must eventually overwhelm the value of c. This guarantees that the next z_n in the sequence will be even larger and the next one larger still — an unbound sequence. Putting it mathematically, for z > 1 and p > 1, it is always the case that

$$z^p > z$$

Since each value must be greater than the previous, the sequence must be unbound once it exceeds the escape radius. In such a case, we know that the number c is not part of the set.

However, for *negative* powers, this rule no longer applies. For a negative power of p, if z_n is large, then the next value z_{n+1} will be *small* because it is a reciprocal. Therefore, just because some value of z_n exceeds the bailout criterion, this does not guarantee that the next value will or that the sequence is unbound. For negative-powered multibrots, the

Figure 7.16

escape time algorithm can result in artifacts: inaccuracies in the resulting map. That is, it may mark a point as not belonging to the set when, in fact, it may actually belong. In a sense, the smoothness of the pebbles in the above images are artifacts of using the escape time algorithm.

This is why some mathematicians and programmers prefer to use a different method for plotting negative-powered multibrots — a more accurate way to discover whether the sequence of z is bound. Several such methods are available and are beyond the scope of this book. The point here is simply that we might consider the above images as *approximations* of a negative multibrot. As mentioned earlier, the bailout criterion affects the size of the pebbles in these maps, and we now understand why — the pebble sizes are an artifact of the bailout criterion. However, the pebble locations are not affected and are therefore intrinsic to the set. We can improve the accuracy of such maps by increasing the escape radius and simultaneously increasing the maximum number of iterations. The results are shown in figure 7.16.

The left panel shows a map of this negative multibrot using an escape radius of 4. The center panel uses an escape radius of 6, and the right panel uses an escape radius of 10. Notice how the pebble size decreases with increasing escape radius. However, the overall shape is relatively unchanged. With increasing escape radius, the largest interior feature approaches the shape of a perfect circle. Several apparent cardioids have also become visible.

Maps of this multibrot using other methods give the same overall shape, and the largest internal features are also the same. But the smooth pebbles may be considered artifacts rather than genuine properties of the set. Alternatively, we could simply adjust the nomenclature. That is, we can *define* this negative multibrot as those points that do not exceed the bailout criterion under the given formula. After all, we are free to define a set by any criterion. And the resulting map is determined by the rules of mathematics. For simplicity, we will continue to use the bailout method to plot maps of the other negative multibrots, using an escape radius of 4.

So, what happens when we consider other negative-powered multibrots? Is there a pattern in the resulting maps as there was with positive-powered multibrots? Figure 7.17 shows the progression as the power of z becomes increasingly negative. We can even begin the sequence with $z^{-1} + c$, as this does result in a fractal, albeit a strangely one-dimensional version.

As with the positive multibrots, there is a logical progression; each multibrot has one more cusp than its predecessor. The number of cusps in a negative multibrot is one greater than the absolute magnitude of the power,

> As with the positive multibrots, there is a logical progression; each multibrot has one more cusp than its predecessor.

MULTIBROTS 113

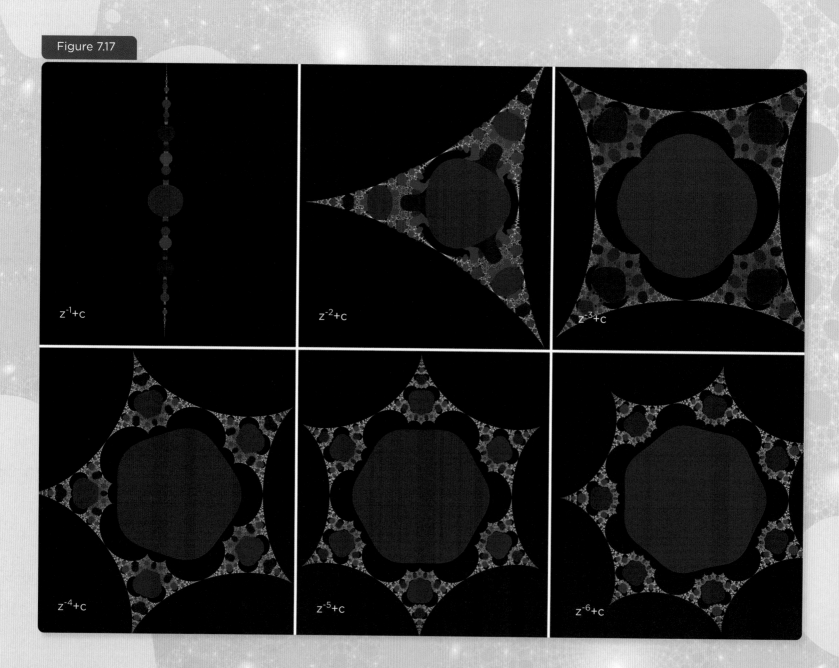

Figure 7.17

whereas the number of cusps in a positive multibrot was one less than its power. Hence, the map $z^{-5} + c$ has six cusps. And so we have discovered a second snowflake fractal, albeit an inverted one since the interior marks points that do not belong to the set.

Zooming in on this negative snowflake multibrot reveals a cobblestone structure similar to the other negative multibrots (figure 7.18). The fractal structure is remarkable and infinite. So each of these negative multibrots contains an infinite degree of complexity on increasingly smaller scales. Each one is a universe that we could spend a lifetime exploring and yet never exhaust its richness. But what does all this mean?

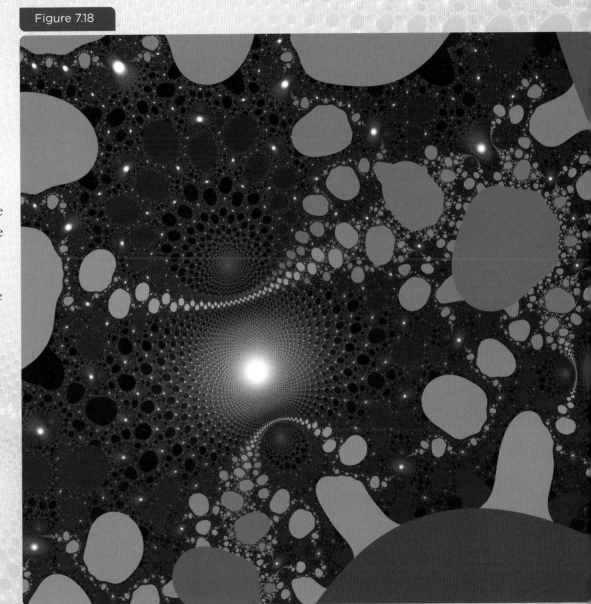

Figure 7.18

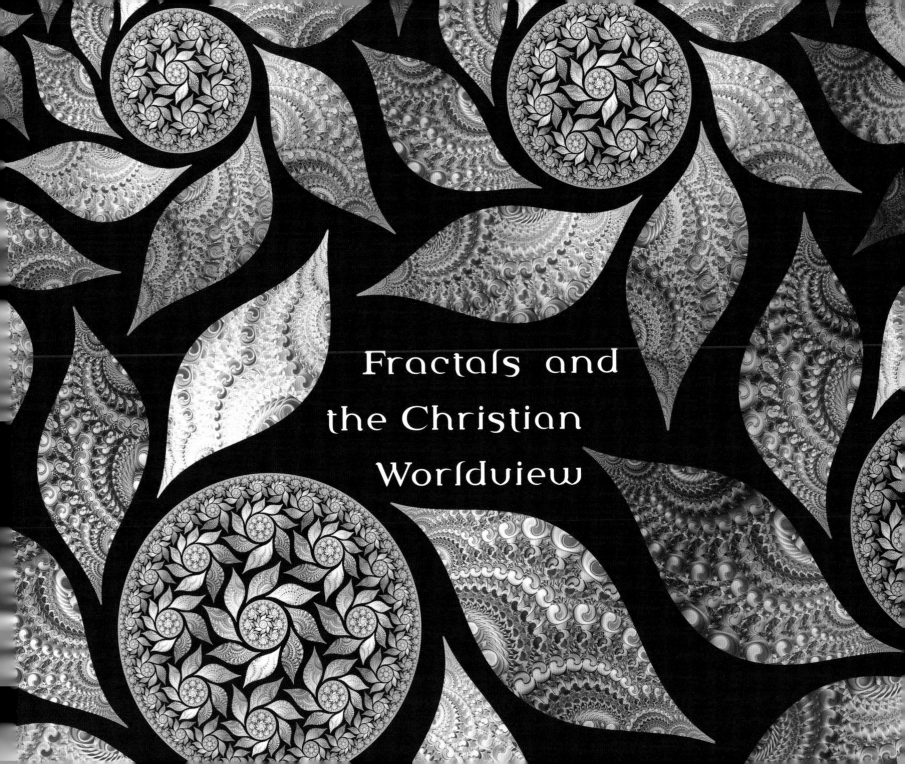

Fractals and the Christian Worldview

We have seen that a secret code of tremendous complexity and beauty resides in mathematics. How can we make sense of this? What is the explanation? I have suggested that the mind of God is ultimately responsible for the infinite beauty and complexity found in the Mandelbrot set. But not everyone agrees with this assessment. Atheists obviously would not appeal to God to make sense of fractals. But can they offer a better explanation? Can any religion other than Christianity make sense of fractals?

Mathematics and the Bible

The Christian worldview — the description of the nature of reality articulated in the Bible — does make sense of fractals. The biblical God is Himself infinite, has a sense of beauty, and His mind determines all truth — including mathematical truths. So, naturally we would expect to find infinite complexity and beauty in mathematical truths.

Furthermore, the Christian worldview can make sense of our ability to discover mathematical truths. The Bible indicates that humans have been made in the image of God — after His likeness. Being made in God's image does not mean that we physically resemble God; God is not a physical being and is normally invisible. Rather, being made in God's image means that we have some characteristics that reflect God's character. Namely, we have the capacity for rational thought. The Bible teaches that God has revealed some of His thoughts to us, and God encourages us to think in a way that is consistent with His character. We can therefore discover some of God's thoughts. Mathematics is one aspect of this — the systematic discovery of how God thinks about numbers.

But how would fractals make sense in an atheistic worldview? And what about non-Christian religions such as Buddhism, Hinduism, or Islam? Can such systems account for the properties of fractals, along with the human mind's capacity do discover such properties?

Many people don't bother to ask these kinds of questions. Like little children, they are content to believe things without considering whether such beliefs are justified and self-consistent. But rationality demands an answer.

Rational people examine their worldview to ensure that it is logical — that it is self-consistent and is backed up by good reasons. Rational people have good reasons for their beliefs, and those reasons do not contradict each other. So, is there a non-Christian explanation for the wonderful truths we have discovered in the previous chapters? Is there a worldview that can make sense of fractals in a self-consistent way? Let's examine some of the answers that people might propose to account for the beauty and complexity in fractals.

THE BEAUTY OF FRACTALS — MANMADE?

Aside from the biblical God, what are some possible explanations for the beauty we have seen in fractals? What is the cause of such beauty? Perhaps the most common non-biblical answer is that human beings are somehow responsible for this incredible beauty. After all, human beings chose which sets to plot — which formulae to investigate. True enough. And human beings chose to plot the map in the Argand plane. Furthermore, human beings chose which colors to use in shading points exterior to the set. These facts are not in dispute. But does this mean that human beings are somehow responsible for the beauty in fractals?

First, while it is true that human beings chose which sets to plot, we have found that beauty exists in just about all the sets we explored. There were a handful of examples that are non-fractal, such as powers of one, zero, and infinity. But literally every other case we tried resulted in a beautiful fractal.

The selected formula doesn't really matter very much. Making adjustments to "c" merely moved, scaled, rotated, or otherwise distorted the shape. And while changing the power of z resulted in new shapes, they were always beautiful.

Furthermore, it is not as if human beings planned the shape of the Mandelbrot set. No man intentionally decided to create a work of art based on a cardioid with infinite circles growing off its perimeter and seahorses, elephants, double spirals, and scepters in its cusps. No man after creating such a work of art then derived a mathematical formula to generate it.

Such a notion just isn't true to the history of the discovery of fractals. Rather, human beings decided to plot a map of which points belonged to a given set and were astonished at the resulting shape. The shape of the Mandelbrot set was not designed by man. It was discovered by man. The fact that human beings were surprised to find such beauty in the Mandelbrot set should refute the notion that human beings created such beauty. You would not be surprised by your own design.

Furthermore, the Mandelbrot set is infinite. It exhibits beauty at all scales, going on forever. But human beings cannot create infinite things because we are finite. There are an infinite number of seahorses in Seahorse Valley. How long would it take a human being to paint an infinite number of seahorses? It would take forever. And yet there are other valleys in the Mandelbrot left to paint. Clearly no human being can create an infinitely complex shape.

Human beings chose to plot the results in the Argand plane. But this is not a "creative" choice. The Argand plane is a very simple and convenient way to plot complex numbers. Although humans arbitrarily selected the x-axis to be the real coordinates and the

> The shape of the Mandelbrot set was not designed by man. It was discovered by man. The fact that human beings were surprised to find such beauty in the Mandelbrot set should refute the notion that human beings created such beauty.

y-axis to be the imaginary, and positive values of x to be to the right and so on, any alternative would simply result in a rotated version of the Mandelbrot set. The beauty would still be present and is therefore not the result of human choice.

But what about the colors? There is no doubt that human beings have selected the lovely color schemes that shade the exterior of the Mandelbrot and other multibrots. Is this color scheme responsible for the beauty in fractals? We can refute such a notion by plotting a fractal in greyscale as in figure 8.1. The fractal is still beautiful even with no color at all. It is the shape of these marvelous plots that is attractive to the eye. And as we said previously, human beings did not design or create the shapes.

A good color scheme does seem to enhance or bring out the beauty in a fractal — but only because the beauty is already there. This is analogous to the way a bit of salt will enhance or bring out the flavor of certain foods. Salt by itself will not create a delicious meal. It can only augment what is already present. Furthermore, the color gradations are assigned on the basis of how quickly the sequence of z exceeds the bailout criterion — which is determined by the rules of math, not human beings. And as a Christian, I would also add that whatever colors we choose to assign to a fractal, we are borrowing these colors from God anyway. (Human beings did not create the properties of light nor the mental perception to the eye's response to combined wavelengths of light that we call "color.") So, while people are responsible for the selection of colors assigned to a fractal, this does not create nor account for the beauty of a fractal. It can only elucidate what is already there.

> So, while people are responsible for the selection of colors assigned to a fractal, this does not create nor account for the beauty of a fractal. It can only elucidate what is already there.

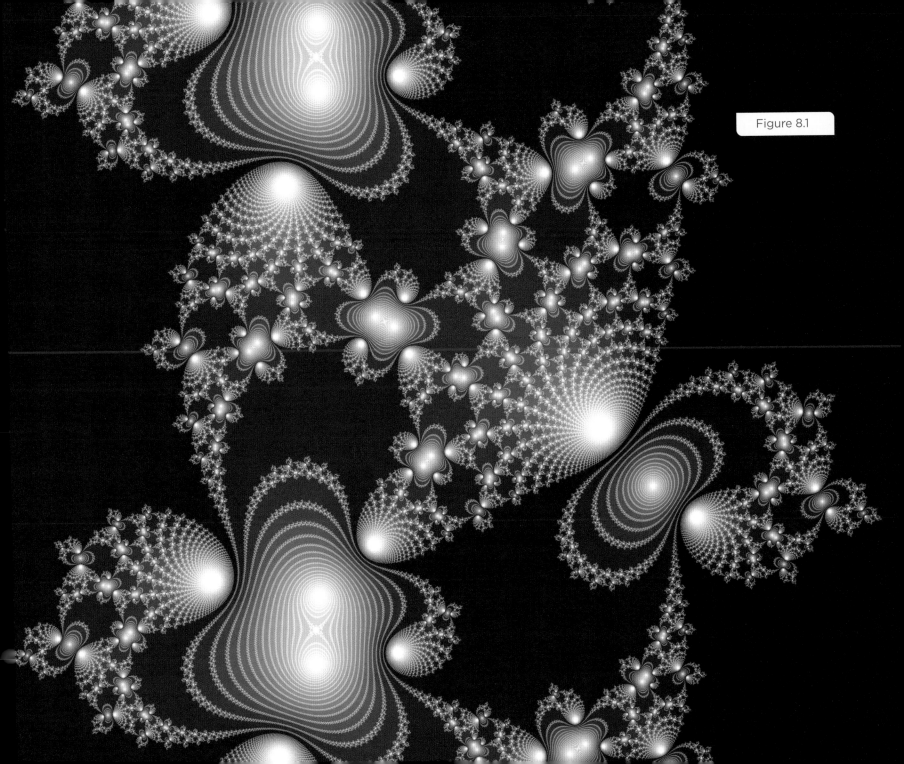
Figure 8.1

The Beauty of Fractals — Computer Generated?

Perhaps computers are responsible for the beauty in fractals. After all, we used computers to plot all the fractal maps seen in this book. So computers created this beauty, correct? Is this a plausible answer?

When the Mandelbrot set was first investigated in the 1980s, some people thought that perhaps the amazing shapes found in plots of this set were artifacts of the computer program — errors introduced by an imperfect algorithm. But we now know that this view is false. The main regions of the Mandelbrot set, such as the main cardioid, the main disk, etc., have been demonstrated to be mathematically true. They are indeed part of the set and not a computer error.

Computers have no creativity at all and therefore cannot be responsible for imparting beauty into the Mandlbrot set. The computer simply follows the rules of mathematics, and the map is the result. In chapter one, we manually analyzed a few values of c to see if they belonged to the Mandelbrot set. We could have done this for many more points and then plotted the Mandelbrot set by hand. If we made no mistakes, we would end up with the same map that the computer did. It would just take a long time.

So, clearly, the computer is not creating the beauty in fractals. It merely reveals what is already there and does so much faster than human beings can do by hand. The computer does not create the beauty in fractals any more than a microscope creates bacteria. The shape of the Mandlebrot set is an inherent property of mathematics. Modern computers simply allow us to investigate that set far more quickly than we can do by hand.

The Beauty of Fractals — Built into Math

Plots of the Mandelbrot set, whether done by computer or by the human hand, are simply the result of following the laws of mathematics for each number in the Argand plane. Therefore, the beauty of fractals comes not from people or from computers, but from numbers and the laws of mathematics. There is an inherent beauty built into numbers and into the rules of mathematics that is revealed when we plot certain sets in the Argand plane. But this just pushes back the question. We must eventually ask why numbers and the relationships between them contain beauty. Furthermore, we have yet to address the question of what causes the complexity in fractals.

The Complexity of Fractals — Manmade?

How do we make sense of the complexity of fractals — the fact that they repeat infinitely at increasingly smaller scales? Fractals have such an intricate and complicated, yet strangely organized, structure. We have seen elaborate tapestries that wind into double spirals, seahorses, elephants, and so forth. What worldview can make sense of this? Such things make sense if mathematics is the study of the way God thinks about numbers. After all, God's mind is infinite, and therefore, He can think of infinite patterns. But many people attempt to explain structures like the Mandelbrot without God. And they tend to fall back on the same explanations we already refuted for the beauty of fractals: namely that either people or computers are responsible for the complexity in fractals.

But the notion that people could create such infinite complexity is absurd on the face of it. No mortal man deliberately planned out the shape of the Mandelbrot set. No one decided to design and create an infinite number of seahorses, elephants, and so on, for no human can create an infinite number of anything. It would take all eternity for a man to imagine each of the seahorses in Seahorse Valley (remember that no two are exactly alike). And then he would have to do it again for the other Seahorse Valley on the opposite cusp. But since he would never finish the first task, he could never begin the second. Nor could he then do this again for each of the infinite number of seahorses in each Seahorse Valley of each mini-Mandelbrot — of which there are an infinite number. Remember, the structures in the Mandelbrot set are not merely infinite, but infinitely infinite!

Man can only plan and build things with finite complexity. But this will not work for the Mandelbrot set or the multibrots. The complexity of just one of these sets exceeds the capacity of the human mind. In fact, it exceeds the combined capacity of all mortal minds — and by an infinite degree.

Of course, human beings defined the set that they intended to investigate. Benoit Mandelbrot of his own volition chose to examine which points were bound under the iteration $z^2 + c$. But he did not plan or design

the map that resulted from this definition. On the contrary, the resulting map was as much a surprise to him as it was to anyone else. Clearly, a person would not be surprised by his own design. Once the set is defined, the resulting map is determined entirely by the rules of mathematics, despite the plans or aspirations of man.

If the Mandelbrot set had been designed by man, then it could have been designed differently. We could have designed it with a valley of penguins and an overall shape that matches the profile of our favorite continent. But this is not the case. You can select the set whose map you want to explore. But the map is determined entirely by mathematics without any human input. We can decide what colors we would like to shade the exterior. But we have no control over the shape itself and its complexities.

The Complexity of Fractals — Computer Generated?

For the same reason, computers are not ultimately responsible for the complexity of fractals because computers do not determine the shape of sets like the Mandelbrot. Rather, computers simply reveal the shape of such sets. The shape of any such set is determined entirely by the laws of mathematics. Computers merely discover this shape more quickly than human beings can do by hand. Nonetheless, the Mandelbrot set could be graphed entirely by hand. It would be tedious. But the same map would result. Computers are simply a tool to probe what already exists in the abstract world of mathematics.

The Complexity of Fractals — Equation Generated?

Perhaps the formula ($z^2 + c$) created the complexity seen in the Mandelbrot set. After all, this formula results in a map of the Mandelbrot set, whereas other formulae with other powers of z result in different sets. Did the formula create all this wonderful complexity?

There is a sense in which this is true: the formula is responsible for the resulting shape. Hence, different formulae have different maps. But the formula merely describes the set under investigation. It doesn't produce the numbers that belong. For example, consider the set of negative numbers. The fact that we can define this set and investigate which numbers belong to it does not create negative numbers. All numbers already exist. And although we can define any set we like by any mathematical criteria, the numbers that do or do not belong are beyond our control. We cannot force the number 5 to belong to the set of negative numbers no matter how much we try. It is excluded by the nature of numbers.

Likewise, the formula for the Mandelbrot set does define which numbers belong. But this does not — by itself — explain the remarkable complexity of the resulting map. We cannot — from the formula alone — discover the map of the Mandelbrot set. We require two other things: we require numbers, and we require the rules of arithmetic. Let me explain:

If we think of $z^2 + c$ as the machine that generates the Mandelbrot map, this machine requires that we feed numbers into it in order to work. And it uses the rules of arithmetic in order to process those numbers. For example, if I give you the formula but stipulate that you are only allowed to consider the number zero and no other number, you will never get a Mandelbrot map. You will conclude that zero does belong to the set, but no map will emerge because you have not been given access to any other number. Furthermore, you will have to use the rules of arithmetic to discover whether even the number zero belongs to the set.

It would seem that the complexity of the multibrot sets is somehow built into numbers and the laws of mathematics. The exact formula really doesn't matter very much, as nearly any multibrot has a map of infinite complexity. And there are many other equations that we will explore later, all of which have maps of infinite fractal complexity. So it cannot be the formula alone that is responsible for such organization. The formula does matter. But it requires numbers as an input, and it uses laws of mathematics to discover which points belong to the set. Somehow, the astonishing complexity of fractals is built into numbers and the relationships between them.

> Somehow, the astonishing complexity of fractals is built into numbers and the relationships between them.

Numbers themselves and the rules that govern the relationships between them are all conceptual. They exist in the mind and are not physical.

What are Numbers?

Some things are just so basic, so fundamental to our thinking, so second nature to us, that they are hard to define. Numbers are such an entity. We use numbers every day. We understand them. And yet they are surprisingly difficult to define. Even various dictionaries do not always have a consistent answer for this. What exactly are numbers? About the best definition I have been able to find in a dictionary is this one: "Numbers are a concept of quantity." Although this may not capture the fullness of the topic under discussion, it is accurate. And therefore, this is the definition we will use here.

Numbers are what we use to conceptualize quantity. As a concept, numbers are abstract. They are detached from the physical objects they describe. Consider for a moment three apples. Apples are tangible, physical things. You can see them. You can touch them. You can even eat them. They are made of atoms and occupy space. Likewise, you can have three bananas. They too are physical. But what do these three apples and three bananas have in common? One thing they have in common is quantity: they both have "three-ness."

The three apples can be touched. But suppose I asked you to touch just the three-ness of the apples without touching the actual apples. You couldn't do it. Three-ness is an abstract, conceptual property of these apples. Although the three apples are physical, three-ness is not. It is one of the conceptual properties shared by the three apples and the three bananas. And yet three-ness is not made of atoms.

But if we eat one apple, two apples will remain. They now have a different quantifier. Likewise, if we eat one banana, two bananas remain — the same quantifier as the apples. Apparently, it makes no difference what objects the number describes; it is a universal principle that $3 - 1 = 2$. But three of what minus one of what? It doesn't matter. We can abstract away the number (the concept) from the physical thing it describes and work purely

in the conceptual realm. And yet the result will have application in the physical universe.

The point here is that while numbers often quantify something that is physical, numbers themselves are not physical. They are conceptual — meaning they exist in the mind. Numbers are the way we think about quantities. Mathematics is the study of the relationships between numbers. As we think about numbers, we note that there are relationships between them — 5 is greater than 2, 6 times 3 is 18, and so on. There are laws of mathematics — rules that govern the relationships between numbers.

Numbers themselves and the rules that govern the relationships between them are all conceptual. They exist in the mind and are not physical. You cannot stub your toe on a law of mathematics or accidentally swallow one. You cannot see a law of mathematics in a telescope or pull one out of the refrigerator. They are abstract concepts. Laws of mathematics are very real and meaningful, of course — it is simply that they are not made of atoms and do not have a particular location in the physical universe. Given that numbers and the laws of mathematics are concepts that exist in the mind, this raises some very serious problems for non-Christian thinkers. For example, where do laws of mathematics come from?

Given that laws of mathematics are mental concepts, secularists will often claim that human beings are responsible for these laws. After all, the human mind can create concepts. The human mind can think mathematically. Secularists often see mathematics as a creation of man. But there are some very serious problems with this view.

Perhaps the most obvious problem is that laws of mathematics existed before people did. As just one brief example, consider the orbits of the planets around the sun. These orbits are governed by Kepler's three laws, so named because they were discovered by the creation scientist Johannes Kepler in the early 1600s. The third of Kepler's laws relates the period of a planet's orbit in years, with the planet's average distance

To do math is to "think God's thoughts after Him."
JOHANNES KEPLER

from the sun in A.U.s. (An A.U. is an "astronomical unit," defined as the earth's average distance from the sun — about 93 million miles.) The relationship is a mathematical one. Namely, $p^2 = a^3$, where p is the period of the planet and a is its average distance from the sun.

Did planets fail to follow this mathematical formula before people came around and created math? Clearly not. Secularists assume that the planets followed this mathematical rule long before people existed. So, clearly, people did not create at least this mathematical rule. Some might respond, "People created mathematics to describe the orderly patterns found in nature, like the orbits of the planets." But this falsely shifts the definition of mathematics to one of mere notation. No doubt, human beings created the notation system we use to describe numbers. But the patterns that exist between numbers — the actual laws of mathematics that the planets obey — existed before people. And it is these laws that we are asking the secularist to explain.

If people created laws of mathematics, then they could have created them differently, just as people invented cars and there are many different types. So, people could have decided that 1 + 1 = 3, or that the square of the hypotenuse of a right triangle is the simple sum of the other two sides rather than the sum of their squares. We could have decided that the circumference of a circle is merely three times its diameter rather than pi, whose decimal expression is infinite and unpredictable (3.14159265 . . .). That would certainly be easier.

People did invent certain notations and definitions used in mathematics. As such, there are some notational differences (compare our Arabic numerals 1, 2, 3, with Roman numerals

Johannes Kepler - Kepler Laws

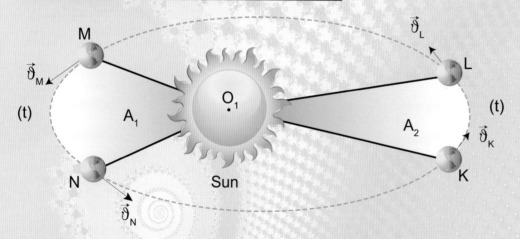

Kepler's second law states that planets sweep equal areas in equal times.

I, II, III). But once the definitions are in place, people have no control over the resulting relationships between numbers. Hence, people did not create laws of mathematics. Rather, people discovered them.

By applying logic to numbers and passing down our findings over time, human beings have discovered many wonderful mathematical truths. But we did not create these truths. To discover something, it must already exist in a form that is "covered" to us so that it may be revealed as we "dis-cover" it. Hence, 2 + 2 equaled 4 before humans existed, and the Pythagorean theorem was true long before Pythagoras proved it. The circumference of a circle has always been its diameter times pi, and humans still don't know the full (infinite) decimal expression of pi. Laws of mathematics do not depend upon people, and they existed before people.

And now we see very clearly the secularist's dilemma. Laws of mathematics are conceptual — like all concepts, they exist only in the mind. Yet, laws of mathematics existed before the human mind. The atheist assumes that the human mind evolved over time and that in the distant past there were no minds in the universe. Yet, he also assumes that the universe obeyed mathematical laws at that time. Being conceptual, mathematical laws require a mind. But the atheist believes that there is no mind governing the universe. The atheistic worldview is inherently contradictory.

Non-Christian religions are similarly stymied by the existence and properties of mathematical truths. The Hindu worldview certainly cannot make sense of mathematics because Hinduism is monistic. A monistic worldview is one that does not accept distinctions: "all is one." The only number that can exist in a Hindu worldview is the number one. No others can exist because they would be distinct, which isn't allowed in Hinduism.

Nor can any polytheistic religion make sense of universal laws of mathematics. If mathematics reflects the thinking of God, then the polytheist will have to ask, "Which god?" In a hypothetical universe ruled by multiple gods, each god could have a different way of thinking about numbers; hence, there would be many different and contradictory sets of mathematics. The Pythagorean theorem could be true in some places at some times, and not true at other places and other times. But this isn't true to the nature of mathematics.

> If people created laws of mathematics, then they could have created them differently. Just as people invented cars, and there are many different types.

The Biblical Worldview and Mathematics

For the Christian, we can have laws of mathematics before people existed because we understand that a Mind has always existed. The universe has obeyed mathematical laws since the instant it was created because God's mind controls the universe. We can explain why laws of mathematics do not change with time — because God doesn't. We can explain why laws of mathematics work everywhere in the universe — because God upholds all creation by the expression of His power (Hebrews 1:3). Mathematics is the way God thinks about numbers. And since His mind determines truth, we must learn to think in a way that is consistent with His character if we are going to do math properly. When we discover a law of mathematics, we have learned something about the way God thinks. To do math is to "think God's thoughts after Him," a phrase written by the 17th-century astronomer Johannes Kepler.

God, as described in the pages of the Bible, makes sense of the properties of mathematics seen in fractals. The Lord has a sense of beauty — we see this in the physical universe. So it stands to reason that we might also find beauty in the abstract world of mathematics where we are learning how God thinks about numbers. Furthermore, God's mind is infinite. And so it is not a problem for God to place an infinite level of complexity into numbers, which is then revealed when we plot maps of certain sets. We don't have to understand everything about fractals in order to make sense of their basic properties. The consistent Christian expects such things, given what the Bible says about the character of God.

> When we discover a law of mathematics, we have learned something about the way God thinks.

No other religion, philosophy, or worldview can make logical sense in a self-consistent way of the evidence we have uncovered. As such, fractals demonstrate the truth of the Christian worldview. As we continue to explore the nature of fractals, we will find even further confirmation of the Bible.

$X_{n+1} = aX_n(1$
$\frac{1}{s^2}\frac{\partial^2 u}{\partial t^2} = \frac{\partial^2 u}{\partial x^2} + \frac{\partial^2 u}{\partial y^2} + \frac{\partial^2 u}{\partial z^2}$
$E = mc^2$
$\log_a(xy) = \log_a x + \log_a y$
$E = mc^2$
$E = mc^2$
$E = mc^2$
$X_{n+1} = aX_n(1 - X_n)$
$a^2 + b^2 = c^2$
$a^2 + b^2 = c^2$

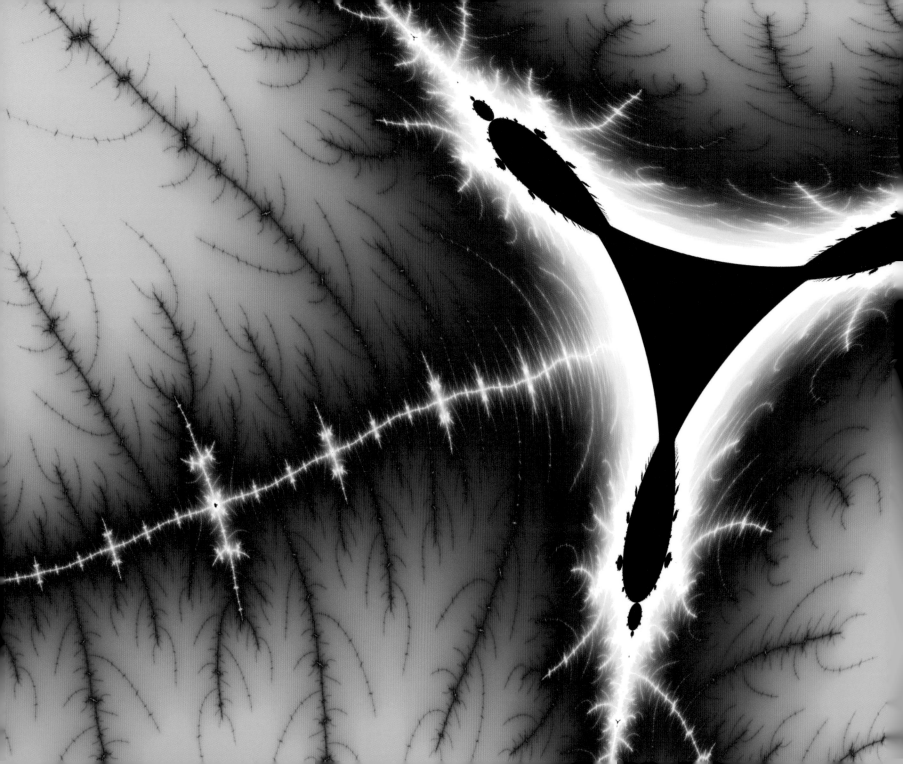

All the fractals we have examined so far are multibrots. They are the plots of sets of all c where z is bounded under the iteration $z^p + c$. And we have seen that the Christian worldview can make sense in principle of the beauty and complexity that results when such sets are plotted in the Argand plane. There is beauty built into math by the Lord, whose mind is ultimately responsible for numbers and all mathematical truths. This being the case, we might expect to find beauty in sets that are not multibrots. We will explore a few such sets in this chapter.

The first set is similar to but distinct from the Mandelbrot set and involves a mathematical concept called the complex conjugate. Fortunately, this concept is very simple; the complex conjugate of a number has the same real value as the number, but the imaginary component has the opposite sign.

For example, consider the complex number 3 + 4i. The real component is 3, and the imaginary component is 4i. So the complex conjugate of this number is 3 − 4i. Likewise, if we consider the number 7 − 4i, then the complex conjugate would be 7 + 4i. If you perform the operation twice (if you take the complex conjugate of a complex conjugate), you get the original number. We denote the complex conjugate of a number z by placing a bar over it: $\bar{z}$. When read aloud, this is pronounced "zee-bar." So, for any number z:

$$z = x + iy$$
$$\bar{z} = x - iy$$

A popular fractal is based on a formula that uses the complex conjugate:

$$\bar{z}_n^2 + c = z_{n+1}$$

Notice that this formula is identical to the Mandelbrot formula except that the z^2 term is replaced by $\bar{z}^2$. That is, we take the complex conjugate of z before squaring it. This small change results in a very interesting and beautiful map shown in figure 9.1.

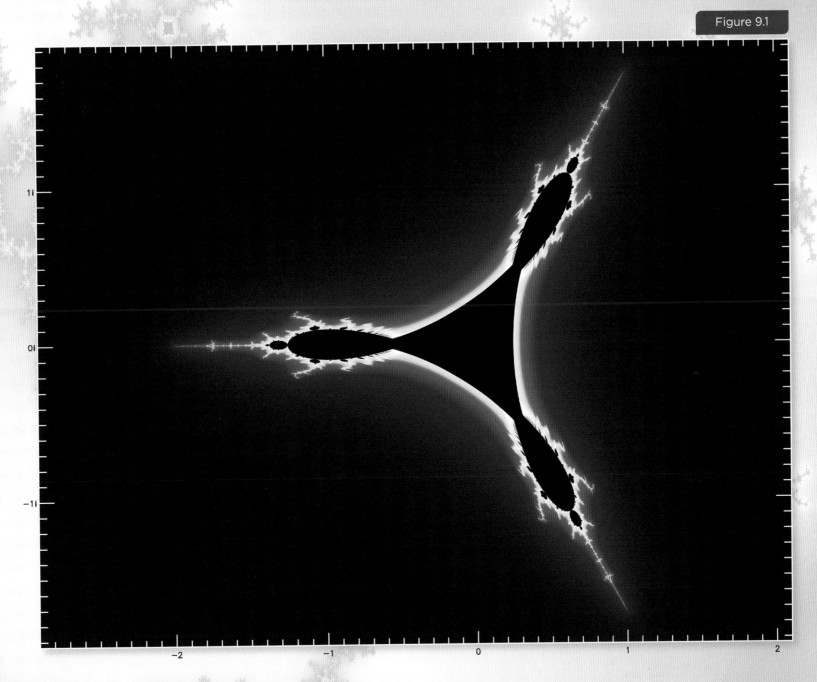
Figure 9.1

This set is called a "tricorn" due to the map's similarity to the hat. It is also called a "mandelbar" due to the fact that it is based on a barred version of the Mandelbrot formula. It has similarities and differences to the Mandelbrot map. The similarities are the many antennae that extend away from the set, the bulbs upon bulbs that occur on the three primary lobes, and the cusps between them. We are not surprised by the convoluted perimeter at the extremities.

However, the central area of a tricorn appears to have a perfectly smooth perimeter — a feature not seen in the Mandelbrot set, which is "wiggly" everywhere. Zooming in on these smooth areas reveals no surprises — merely a sharp and smooth demarcation between those points that belong to the set and those that do not. The shape also has a perfect three-fold symmetry.

Whereas the Mandelbrot set has one primary antenna extending to the left, the tricorn has three; one straight to the left, one on the upper right, and one on the lower right. We zoom in on the left antenna in figure 9.2. We readily see the lightning bolt patterns extending away from the bulbs on the right side of the plot. Notice also that the bulbs appear stretched; they are ellipses rather than the circles we saw on the Mandelbrot. We will find that this distortion-by-stretching is a common feature of the tricorn.

On the left side of the plot in the white box, we are pleased to see that the tricorn has a mini-tricorn growing off the antenna. The tricorn is a fractal! The tricorn has three primary minis of equal size, one sprouting off each of the three main antennae. Figure 9.3 is an enlarged plot of this mini. But notice that the mini appears stretched relative to the original in the direction of the antenna. Otherwise, the mini is basically a copy of the original, except for extra dendrites that extend away from its perimeter, just as we saw in the mini-Mandelbrot.

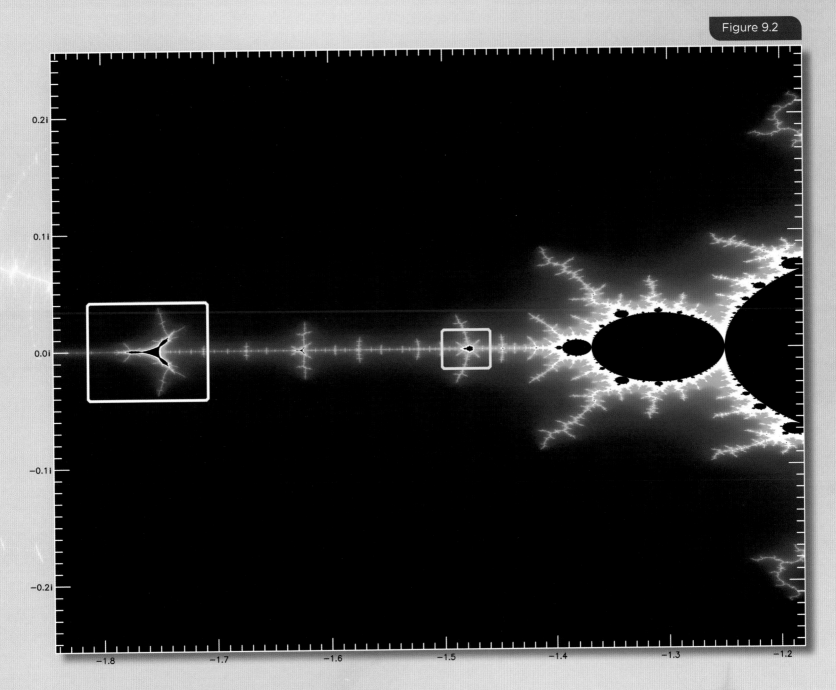

Figure 9.2

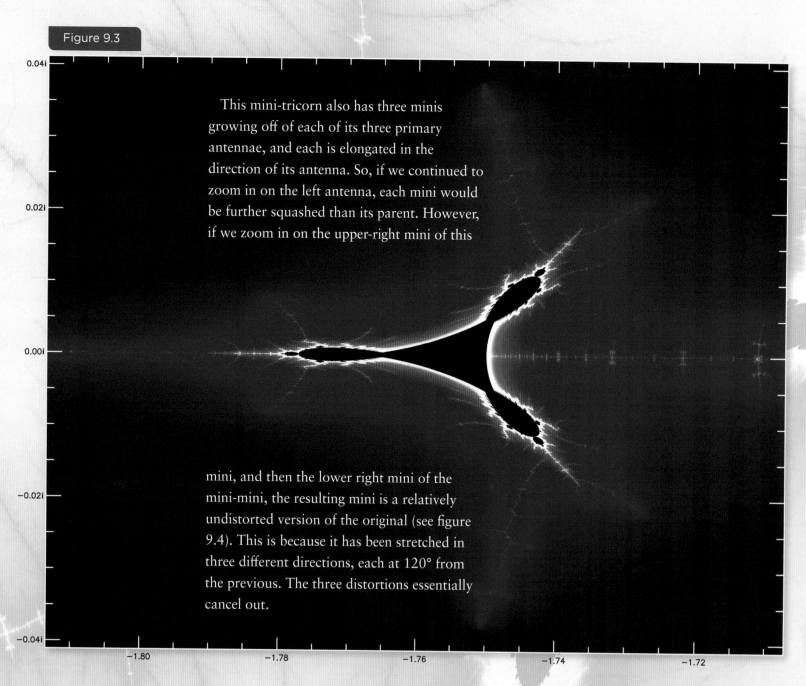

Figure 9.3

This mini-tricorn also has three minis growing off of each of its three primary antennae, and each is elongated in the direction of its antenna. So, if we continued to zoom in on the left antenna, each mini would be further squashed than its parent. However, if we zoom in on the upper-right mini of this mini, and then the lower right mini of the mini-mini, the resulting mini is a relatively undistorted version of the original (see figure 9.4). This is because it has been stretched in three different directions, each at 120° from the previous. The three distortions essentially cancel out.

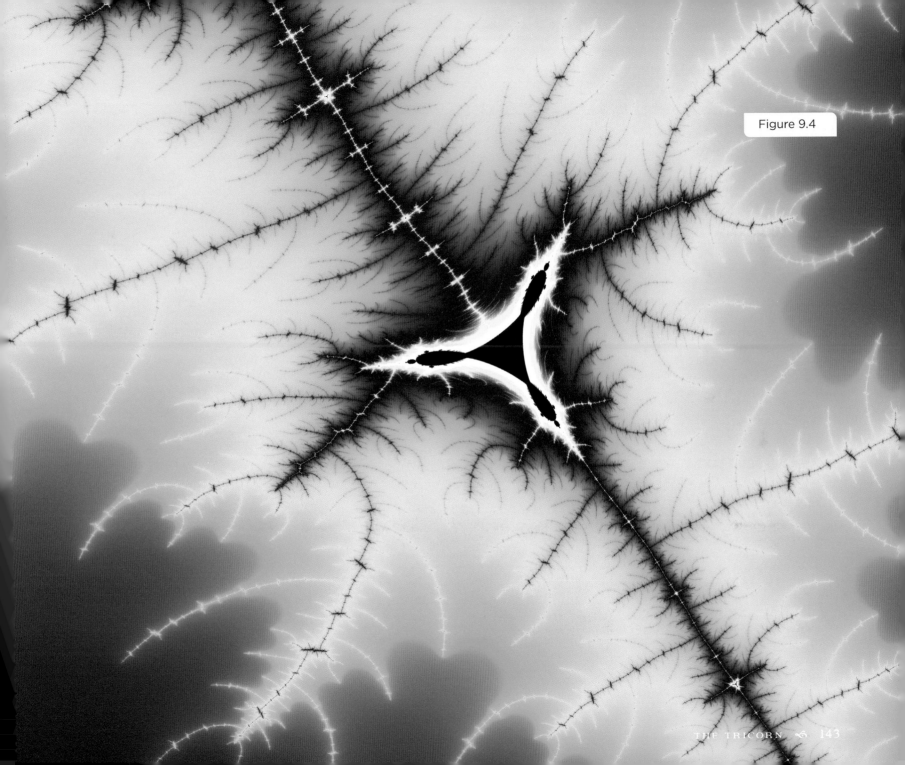
Figure 9.4

A Trojan Mandelbrot

Returning to figure 9.2, we again examine the leftmost antenna of the tricorn. We saw previously that this antenna contains a miniature version of the entire tricorn. But it also contains another delightful surprise. We zoom in on the yellow boxed region to reveal figure 9.5. Amazingly, we find not a mini tricorn, but a mini Mandelbrot!

This may be surprising because the equation for the tricorn is fundamentally different from the Mandelbrot, and yet the Mandelbrot shape is found within its map. It seems that the shape of the Mandelbrot set is a fundamental property of numbers and can appear even in non-Mandelbrot equations.

In fact, the tricorn not only contains an infinite number of mini-tricorns, but also an infinite number of mini-Mandelbrots. Along each antenna are an infinite number of buds, each of which is either a mini-tricorn or a mini-Mandelbrot. We note that the miniature version seen in figure 9.5 is distorted from the original Mandelbrot, appearing stretched along the x-axis just as the mini-tricorn was.

This common property of the tricorn in which miniature versions appear stretched often gives them the appearance of being viewed in perspective, as if plotted on a surface that is tilted from our point of view. This is the case even with deeper plots that reveal the intricate internal structure of this fractal (see figure 9.6).

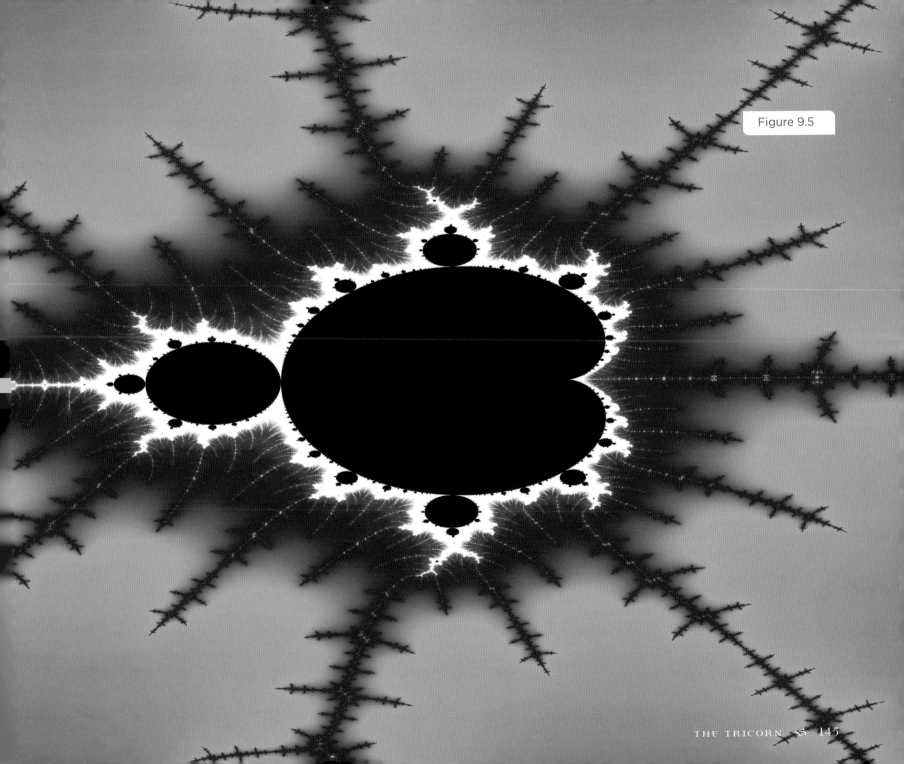

Figure 9.5

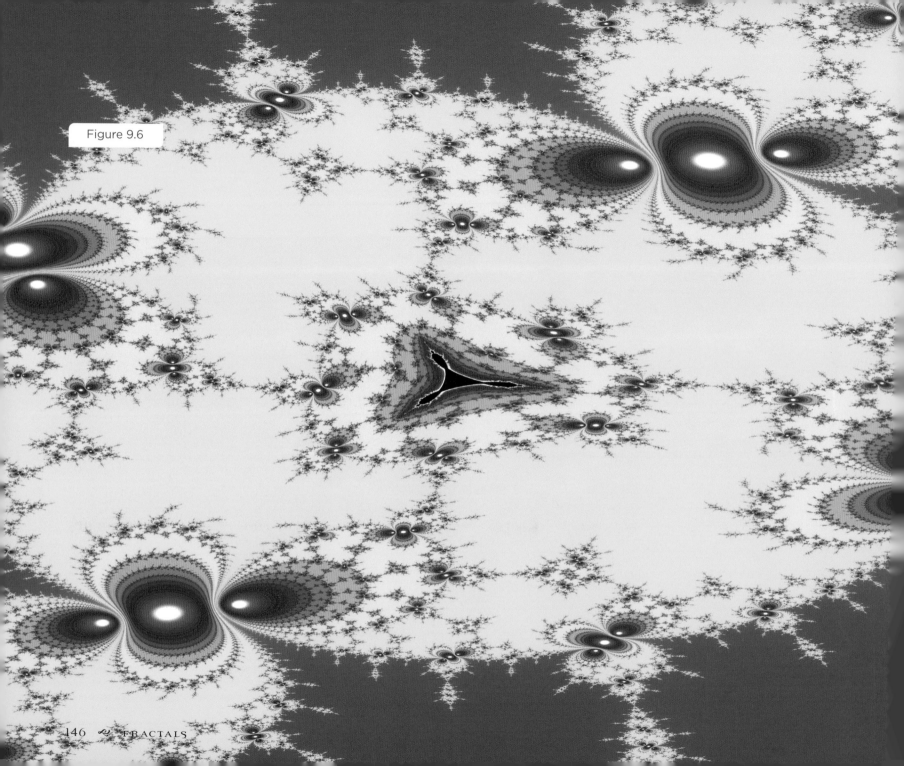
Figure 9.6

Multicorns

Since it is an infinitely complex shape, we could spend all eternity exploring the map of the tricorn. However, if we change the power of z in the formula, we end up with a different shape. These variations are sometimes called muticorns. Just as the multibrot sets are higher-powered versions of the Mandelbrot set, so the multicorns are higher-powered versions of the tricorn. And we find that there is a logical progression in the resulting shape as we increase the power of z. Several multicorns are plotted in figure 9.7.

Notice that each multicorn has one additional "arm" from the previous. Furthermore, the number of arms is simply the power of z plus one. Hence, the multicorn formula $\bar{z}^5 + c$ results in snowflakes with six-fold symmetry. Multicorns are fractal, so the snowflake multicorn has many smaller versions of itself built in — see figure 9.8. Again, the stretching effect often makes the miniature versions appear as if in perspective.

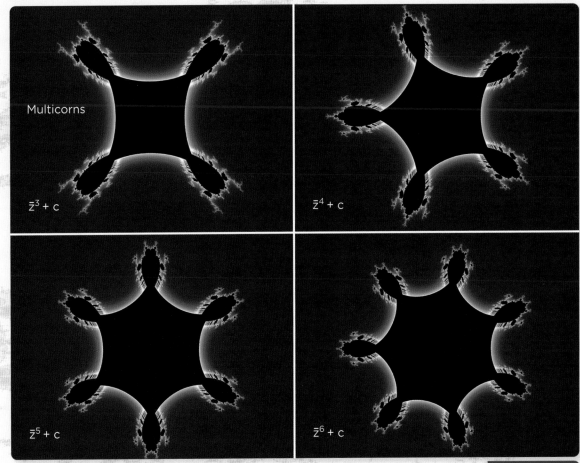

Figure 9.7

Figure 9.8

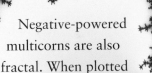

In addition to miniature multicorns, miniature versions of the multibrot sets are also found within the map of a given multicorn. The power of the mini multibrot matches the power of the multicorn. So, for example, the fifth power "snowflake" multicorn also contains fifth-power multibrots, which are four-lobed. Some of these are visible in figure 9.8.

Negative-powered multicorns are also fractal. When plotted using the escape time algorithm, they exhibit the same type of pebble structures seen in negative-powered multibrots, but the overall shape is quite different. So we again confirm that the beauty and complexity of fractals is not something limited to a particular equation that human beings invented. Such characteristics are an inherent property of mathematics that stems from the mind of God.

The Sine Function

In addition to the equations we have covered so far, many other fractals are found in the plotted sets of points bound under iteration. Such sets are often based on a polynomial. A polynomial is any combination of terms of the form Az^p where z is the variable, A is a constant, and p is the power. For example, $2z^3 + 5z^2 + 3z$ is a polynomial. When we consider those points bound under iteration for a given polynomial, we find that the resulting sets are often fractal. There are an unlimited number of fractal polynomial sets, each with infinite complexity comparable to the Mandelbrot.

Trigonometric Functions

However, some mathematical functions cannot be exactly expressed as a finite-term polynomial. Trigonometric functions are such examples. Among other things, they are used to convert between angles and segment ratios in a right triangle. Three of the most common trigonometric functions are the sine, cosine, and tangent functions. Less commonly used (because they are merely reciprocals of the previous three) are secant, cosecant, and cotangent.

Consider the sine function. For a given angle, θ (theta), the sine of θ is the ratio of the segment opposite the angle to the hypotenuse of that triangle (see figure 10.1). It also expresses the y-position of a point in a unit circle at angle θ. For real values of θ, sin(θ) has a maximum value of 1 and a minimum value of -1.

In lower-level mathematics, angles are often measured in degrees, with ninety degrees representing a right angle. But in more advanced mathematics, we generally use radians instead of degrees. The number of radians in a right angle is exactly half of pi, or approximately 1.5707963. Radians are a more natural unit of angle, for many reasons. One is because an arc of a unit circle will have a length equal to its angle in radians. Just as there are 360 degrees in a circle, there are 2π radians in a circle. You can convert from degrees to radians by dividing by 180 and multiplying by pi. In this book, we will use radians exclusively.

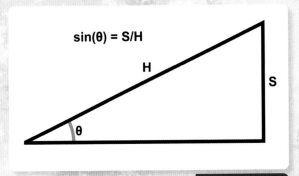

Figure 10.1

Euler's Identity

Another function often used in introductory calculus is the exponential function: e^x. Here, e is "the natural number" — an irrational constant (like π) that has an infinite, non-repeating decimal expansion, which begins 2.718281828459045. The practical applications of the exponential function are numerous. Many forms of decay (such as radioactive decay) or growth (such as bacterial growth) are described by an exponential function.

When considering only the real numbers, there is no obvious connection between the exponential function and the trigonometric functions. But the 18th-century mathematician Leonhard Euler (pronounced OY-ler) found that these functions are related when we consider complex numbers. In particular, he discovered what we now refer to as Euler's identity:

$$e^{ix} = \cos(x) + i\sin(x)$$

When we set $x = \pi$, we discover an amazing result:

$$e^{i\pi} = -1$$

This simple formula is often considered the most remarkable mathematical discovery of the 18th century because it showed that four basic constants (e, i, π, -1) are related in a way that no one expected. But for our purposes, the Euler Identity allows us to explore complex sets that include trigonometric functions.

We understand from high school math what the "sine of theta" means when theta (θ) is a real number: the ratio of the opposite leg to the hypotenuse in a right triangle. But what if θ is imaginary? What if θ is complex? Using the Euler identity and some algebra, we can express the sine of a complex number (z) as follows:

$$\sin(z) = \frac{e^{iz} - e^{-iz}}{2i}$$

So the sine function is perfectly well defined even for complex numbers.

The Sine Function

Now, this is all background information, and we need not concern ourselves with the details of these formulae. The point is we can program a computer to compute the sine of any complex number. We could do the same for the other trigonometric functions. In this chapter, we shall explore the set of all points bound under the sine function iteration:

$$\sin(z_n) + c = z_{n+1}$$

This is just like the Mandelbrot set except, instead of evaluating the function z^2, we are evaluating the function $\sin(z)$. We need not actually know much about what the sine function is in order to appreciate the beauty of this set. For our purposes, the above formula is just one function of an infinity of functions that we are exploring to see what shape emerges in the complex plane. Will it be fractal?

The result is shown in figure 10.2. Black regions represent points that are bound under the iteration $\sin(z) + c$. As before, colored regions indicate unbound points, where the shading indicates the speed with which z escapes to infinity under the escape time algorithm. We cannot plot the entire set because it has infinite area: its height is finite, but its width extends to infinity. However, we don't need to see any more than is plotted in figure 10.2 because the shape repeats exactly along the x-axis. This is because the sine function is periodic; namely, $\sin(z) = \sin(z + 2\pi)$ for all z. This is the first periodic function we have explored.

We see some features common to the Mandelbrot set. There are several cusps whose valleys contain interesting and beautiful substructure. As we zoom in, we will find double spirals, but they have a different internal texture than we have observed so far. We see an elaborate tapestry of geometric shapes extending from the exterior, as with many of the miniature versions of the Mandelbrot.

But we also see significant differences. There are no obvious antennae or dendrites on this map. And the elephant and seahorse structures of the multibrots are not apparent

Figure 10.2

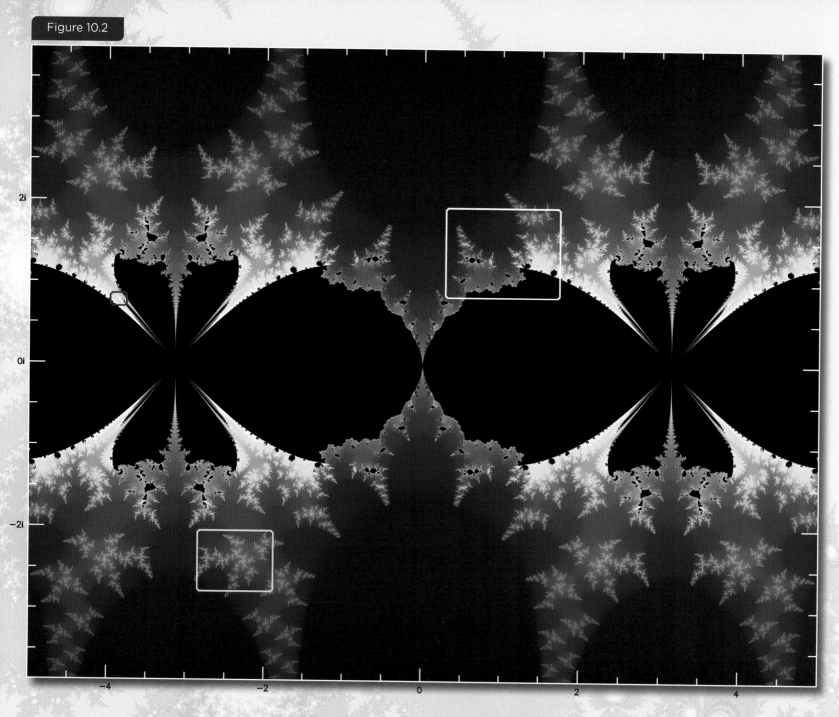

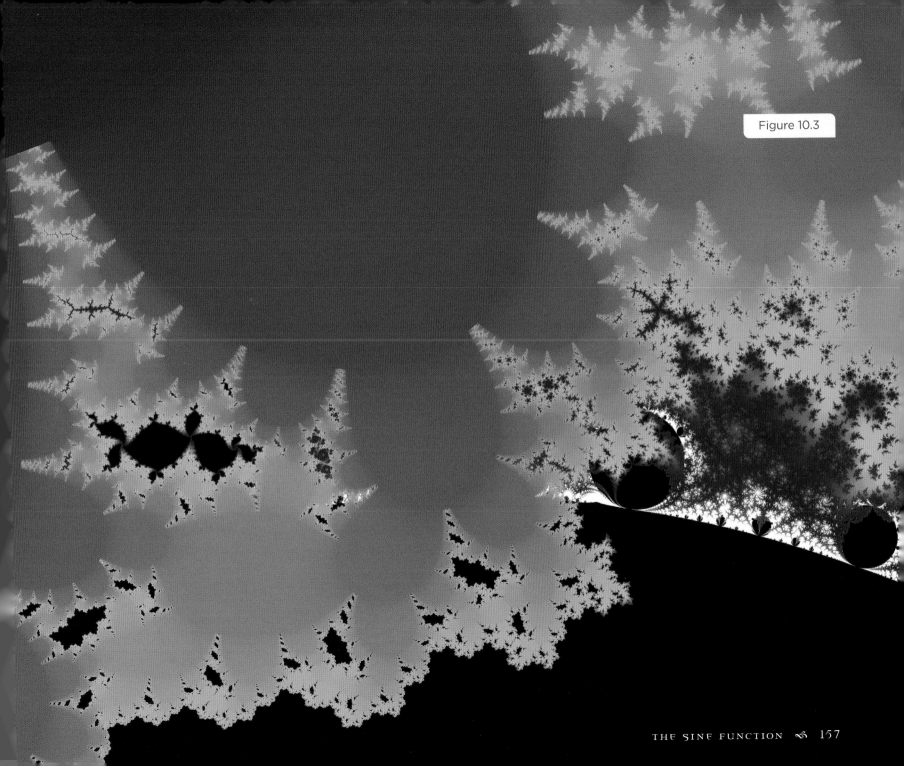
Figure 10.3

here. Furthermore, even at this scale, we can see a number of "islands" — regions belonging to the set that are separated from other regions. The sine set is not connected, whereas the Mandelbrot set is.

The white boxed region is displayed with greater magnification in figure 10.3. Here we clearly see that the "islands" are truly separated from the "mainland" with no connecting structures. This set has a broken appearance, as if smaller pieces were breaking off of a much larger glacier. We also see structures somewhat resembling tiny Christmas trees above and surrounding the islands. We will find that such "trees" are a common structure in the sine map.

On the lower right of the image, we see hints of the circles that grew along the perimeter of the Mandelbrot set. But here they appear damaged, almost as if two different overlapping fractals were fighting for dominance. Above and to the right of the largest quasi-circle is a lovely open spiral. We zoom in on this feature in the next figure.

In figure 10.4, we see that the spiral is composed of many smaller spirals that are not connected to each other. However, each resembles the larger spiral, which illustrates scale-invariance. So this set is indeed fractal, even though the overall shape of the set is not found within itself. The other noteworthy aspect of this spiral, aside from its astonishing beauty, is that it resembles a particular type of spiral galaxy known as a flocculant. These also have bits of apparently disconnected material, which can be traced into a larger spiral structure.

Returning to figure 10.2, we now zoom in on the red box to see the substructure contained within the valley. The result is figure 10.5. Note that unlike the valleys in the Mandelbrot, this valley has a "peninsula" within it, extending from the lower right side. On the lower left, we again see hints of the circles that grew along the valleys of the Mandelbrot, but here they are interrupted by strands that appear almost as if they have been painted over a background.

Figure 10.4

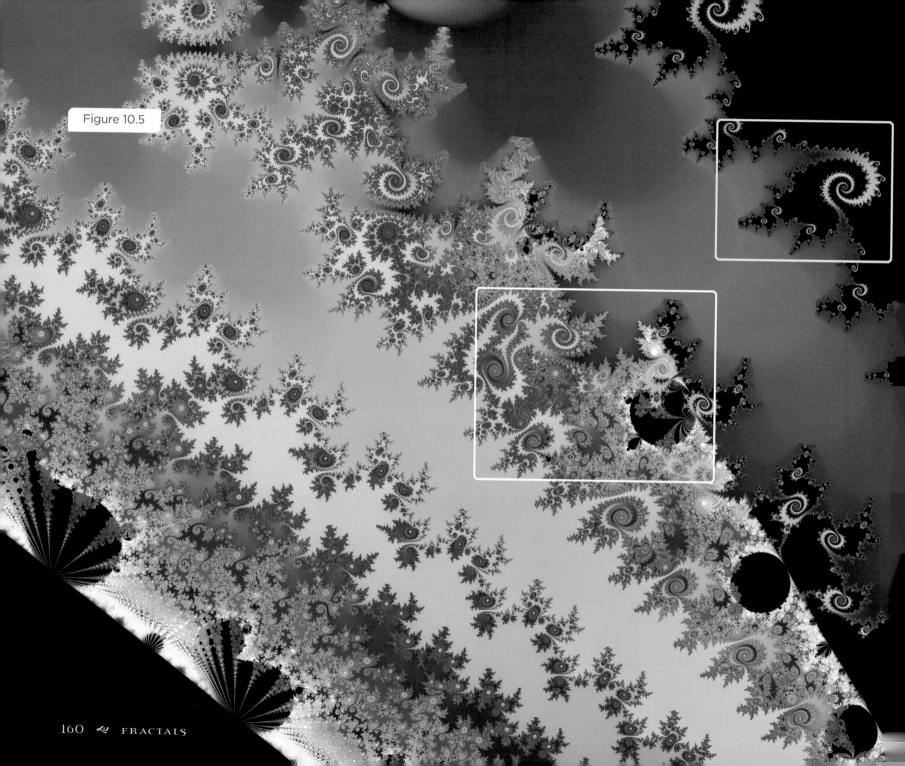

Figure 10.5

Figure 10.6

Zooming in on the yellow box, we go to figure 10.6. This is a double spiral, but it is inverted relative to the double spirals found in the Mandelbrot set. That is, the exterior is black — points that do belong to the set — while the narrow spiral strands themselves are empty. Notice that the perimeter of the spiral has a sawtooth appearance owing to numerous small Christmas tree structures. Ubiquitous tiny versions of this double spiral are found along the perimeter of the large one. The fractal nature is stunning.

Returning to figure 10.5, we now zoom in on the white boxed region resulting in figure 10.7. We again get the impression that we are looking at two overlapping fractals, particularly on the right side of the plot. The mind readily perceives a large black circle that appears partially eclipsed by blue gossamer. Some of the gossamer structure seems to change color from yellow to blue as we go from the outside to the inside of the circle. Various double spirals are visible throughout the plot but with a sawtooth or zipper appearance due to the tiny Christmas tree structures along their edges.

An enlargement of the center of this region is shown in figure 10.8. We are treated to a gorgeous double spiral just left of center, which is again decorated with tree-like structures along its periphery. In the lower right, we find a web with multiple double spirals embedded within. Like the other fractals we have explored, the sine function exhibits tremendous beauty and apparently continues inward infinitely.

Returning to figure 10.2, we now zoom in on the region within the yellow box. The result, shown in figure 10.9, is a frost-like structure that repeats infinitely. When we zoom in on the white box of this region, the result (figure 10.10) resembles the previous figure. This frost pattern continues infinitely on increasingly smaller scales.

We again return to figure 10.2 and examine a very small region, deep in the first upper valley of the main lobe just left of center. The result is displayed in figure 10.11. We see elaborate bands of gossamer stretching from the lower left to the upper right. The bands are parallel and become closer together as we approach the surface of the set on the lower right. The effect somewhat resembles a sunset.

Zooming in on the boxed region, we obtain figure 10.12. We again see numerous double spirals, single spirals, and Christmas tree structures extending away from them. Yet, all the graphs in this chapter are simply small samples of an infinitely intricate structure. We could spend the rest of our lives exploring its beauty without ever exhausting its riches.

Figure 10.7

Figure 10.8

Figure 10.9

THE SINE FUNCTION 165

Figure 10.10

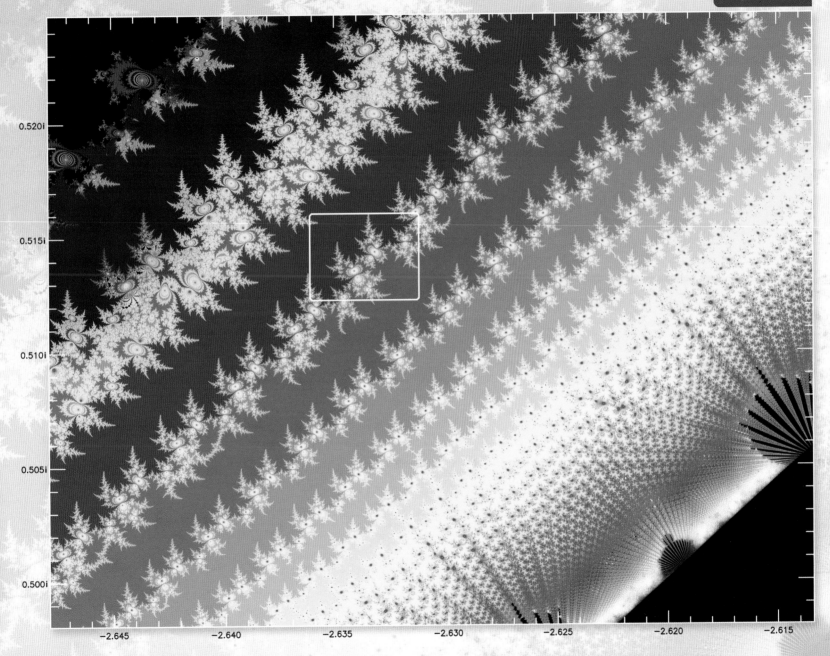

Figure 10.11

THE SINE FUNCTION ~ 167

Figure 10.12

Other Trigonometric Functions

You may be wondering, "What about cosine? What shape do we get when we examine those numbers bound under the formula $\cos(z) + c$?" But, in fact, you have already seen it. The shape of the map of $\cos(z) + c$ is identical to $\sin(z) + c$. The only difference is that the cosine map is shifted leftward relative to sine by a distance of $\pi/2$. This is because $\sin(z) = \cos(z - \pi/2)$.

The map of numbers bound under $\tan(z) + c$ turns out to be rather uninteresting. Its map is entirely black and therefore non-fractal, except for a dusting of points along the x-axis. However, the set of points bound under the exponential function $e^z + c$ is fractal, as shown in figure 10.13.

The exponential set is periodic along the *vertical* axis, again showing the connection between the exponential and trigonometric functions when complex numbers are involved. However, the set is not well-mapped using the escape time algorithm. So, figure 10.13 should be considered an approximation. Although this shape also merits a lifetime of exploration, we now move on to other types of fractals.

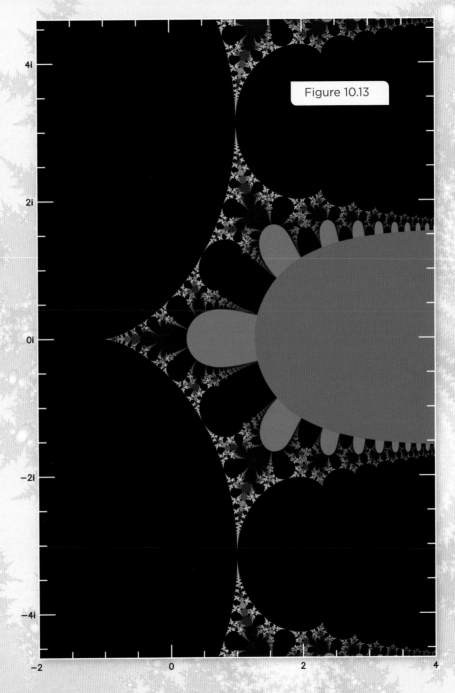

Figure 10.13

THE SINE FUNCTION 169

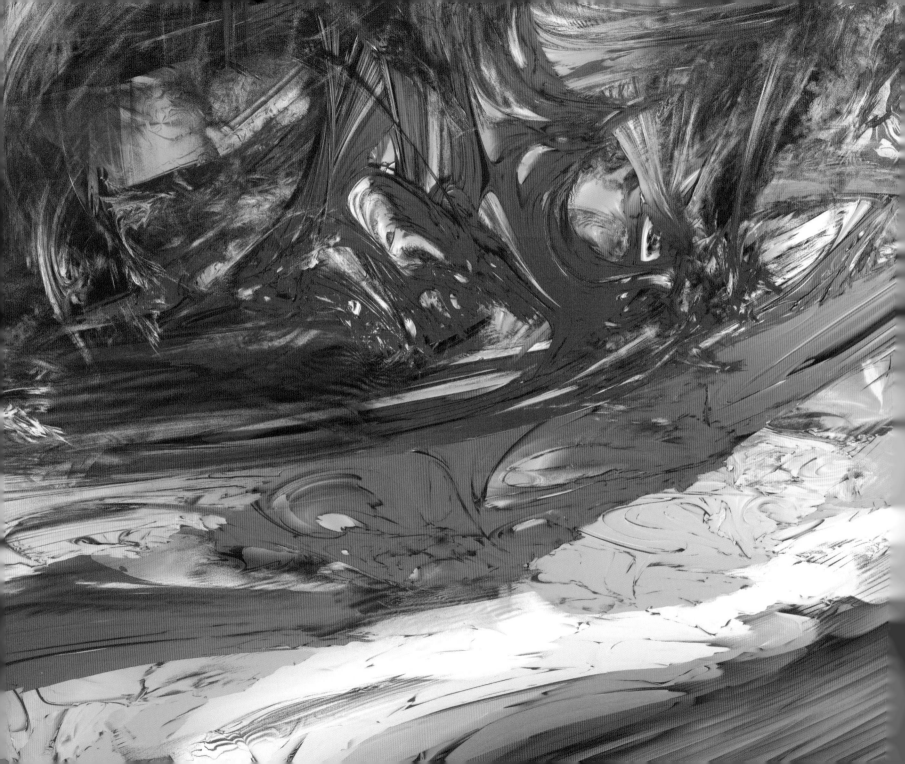

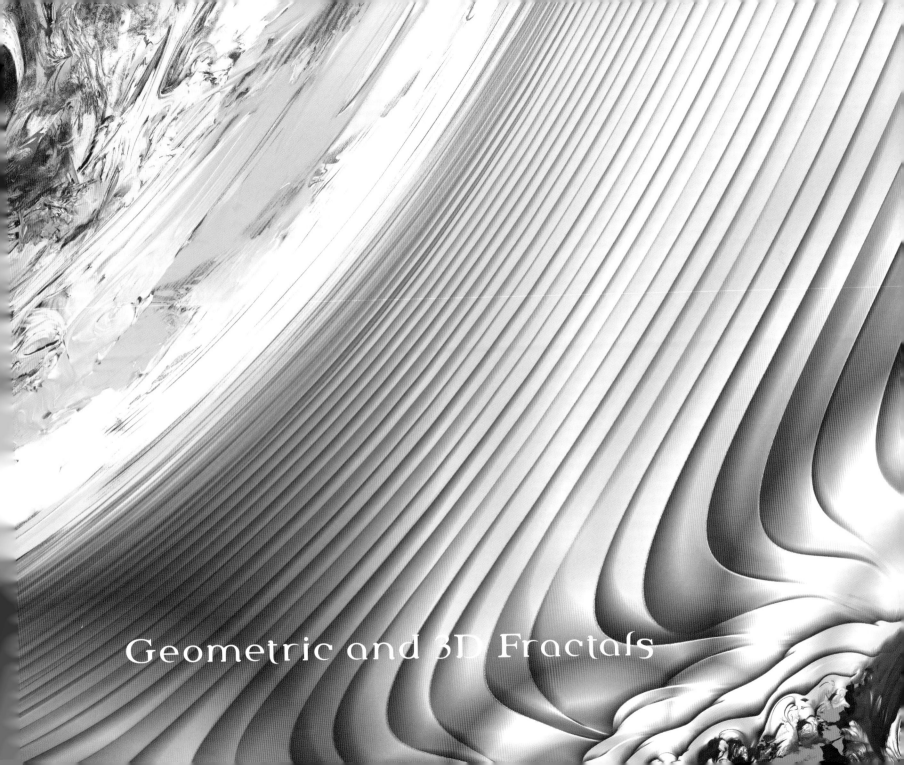

The kinds of fractals we have explored so far have been algebraically defined then geometrically explored. That is, we have defined an algorithm and then mapped the points that belong to that set in the Argand plane. We have been pleasantly surprised by the geometric shapes that result when applying God's rules of mathematics to simply defined algorithms.

However, we can also do essentially the reverse. We can define a fractal geometrically and then discover its algebraic properties. We begin by exploring a geometrically defined fractal known as a Koch (pronounced like "cook") snowflake.

To generate a Koch snowflake, we begin with an equilateral triangle (figure 11.1A). For each of the three line segments, we add a triangle to the exterior, one-third the length of the original (figure 11.1B). This results in a star of David. For each of the remaining twelve line segments, we attach a triangle one-third the length of each segment as before (figure 11.1C). The resulting shape is now beginning to resemble a snowflake.

We continue this process forever, adding a new external triangle one-third the length of each remaining line segment. The final shape resembles figure 11.1D. However, this figure is only an approximation because the perimeter of a true Koch snowflake is infinitely wiggly.

The Koch snowflake has some very interesting properties. Like the Mandelbrot set, the Koch snowflake has an infinite perimeter but a finite area. The total length of all the infinite line segments is infinite, but the total area of the infinite number of triangles is finite. However, unlike the Mandelbrot set, we can compute the exact area of a Koch snowflake. The area is exactly 8/5 times the area of the original triangle.

The way we discover the area of a Koch snowflake involves a branch of calculus dealing with the concept of an "infinite series." It is a truly fascinating topic, the details of which are beyond the scope of this book. However, because God's mind is infinite and because He has graciously revealed Himself to us in various ways, including mathematics, we can actually discover some

Figure 11.1

GEOMETRIC AND 3D FRACTALS ~ 173

truths that involve the infinite! Yes, calculus allows us to conceptually add an infinite number of things and discover the answer in a finite amount of time.

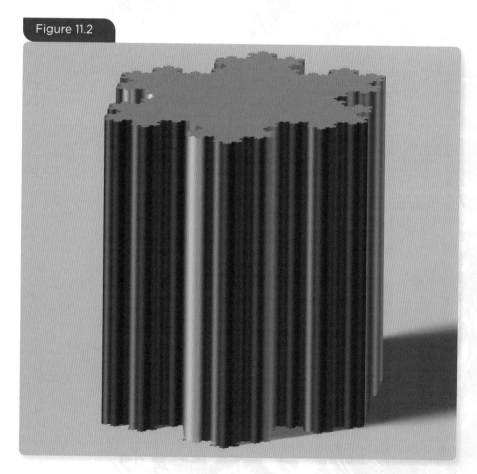

Figure 11.2

The strangeness of a finite area bounded by an infinite perimeter can be illustrated by adding a third dimension. Imagine a cylindric version of the Koch snowflake, one that has been extruded out of its plane. This is like a paint can that looks like a Koch snowflake from the top instead of a circle (see figure 11.2). This paint can would have a finite volume but an *infinite* surface area. In other words, it could not possibly hold enough paint to paint itself! It could not even paint itself on the inside, despite its finite volume. Infinity is a wonderful but counterintuitive truth.

The Koch snowflake, like all fractals, is scale invariant. By zooming in on any section of its perimeter, the pattern resembles any other section of its perimeter. Whether we zoom in one thousand times or ten billion times, the pattern looks basically the same. The Koch snowflake is related to the Koch curve, which is the same concept starting with one line segment rather than a triangle.

Another geometrically defined fractal is the Sierpinski triangle. Like the Koch snowflake, we start with an equilateral triangle. But rather than adding triangles to the exterior, we remove a triangle from the interior that is vertically inverted and with a width one-half that of the original. The result is an area of three triangles, each of which is one-fourth the area of the original (see figure 11.3 top panel). We then do the same thing for each remaining triangle — remove a vertically inverted triangle from the center one-half the linear scale of each triangle. This process continues forever. The resulting ghostly figure is presented in figure 11.3, bottom panel.

Since at each iteration we are removing a triangle one-fourth the size of each remaining triangle, the area of each iteration is ¾ the area of the previous iteration. Since this continues indefinitely, the total area of the complete Sierpinski triangle is exactly zero. Strangely, there are points that belong to the Sierpinski triangle at all iterations, such as the points on the vertices of any (black) triangle. And since there are an infinite number of

Figure 11.3

A Sierpinski triangle

GEOMETRIC AND 3D FRACTALS 175

Figure 11.4

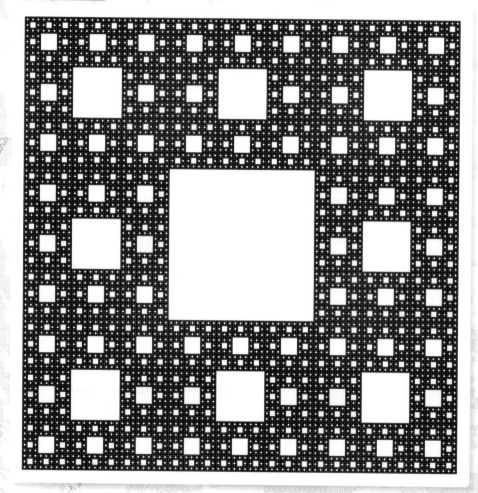

A Sierpinski carpet

triangles, there are an infinite number of points that belong to the Sierpinksi triangle. Yet, their total area is zero!

A similar iteration can be done with squares, in which a square one-third the width of the original is removed from the center. This is repeated forever for each remaining square, and the result is called a Sierpinski carpet (see figure 11.4). Like the triangle, the Sierpinski carpet has a total area of zero, yet there are an infinite number of points that belong to the set.

Since the above fractals are defined by iteration of geometric shapes, we can also apply such reasoning to three-dimensional solids. One example is a Sierpinski tetrahedron (figure 11.5). Rather than a triangle, we begin with a tetrahedron (a "pyramid" with a triangular base). This is replaced by four tetrahedra, each of which has half the width of the original. The same is done for each resulting tetrahedron and so on *ad infinitum*.

The Sierpinski tetrahedron has some strange and wonderful properties. Curiously, the combined surface area of all the mini-tetrahedra at any iteration is exactly the same as the surface area of the original tetrahedron. The volume, however, decreases with each iteration, such that it goes to zero as the iterations go to infinity. So we have the bizarre situation of a shape with finite surface area and yet no volume. Such curiosities are common with fractals.

A similar iteration can be done with cubes. A cube can be divided into 27 smaller cubes of equal size, resembling a Rubik's cube. Remove the central cube on each of the six faces, along with the center cube. Do the same for each remaining cube and continue forever. The result is sometimes called a Sierpinski cube but is more commonly known as a Menger sponge (see figure 11.6). Every cube within the Menger sponge is scaled down but otherwise identical to the entire Menger sponge. Notice that each surface of the Menger sponge is a Sierpinski carpet (figure 11.4). The Menger sponge has a total volume of zero and yet has infinite surface area.

Figure 11.5

A Sierpinski tetrahedron after three iterations

GEOMETRIC AND 3D FRACTALS 177

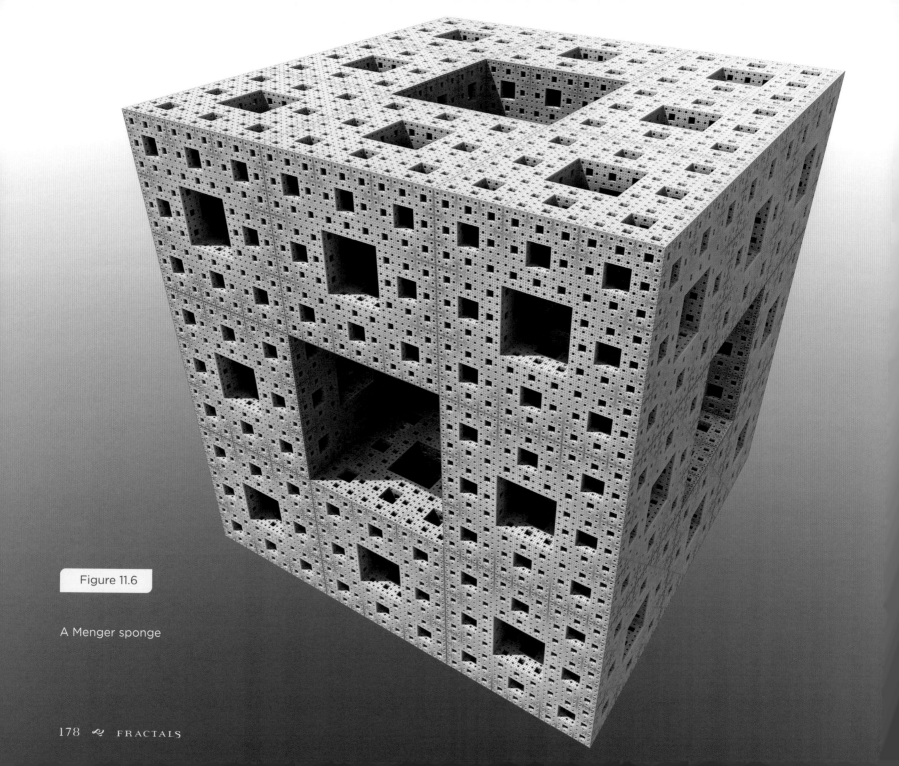

Figure 11.6

A Menger sponge

The Two-Dimensional Nature of Numbers

The Sierpinski tetrahedron and Menger sponge are the first 3D fractals we have explored in this book, in that each has height, width, and depth. And many other three-dimensional fractals are possible when defined by *iteration of geometric solids*. But all the fractals we explored in previous chapters — those defined by *algebraic iterations* — were two-dimensional. They have height and width but no depth because they live in the Argand plane. This prompts us to ask, "Do any *algebraic sets* have a three-dimensional fractal map?"

There are techniques to give a type of "depth" to the Mandelbrot set. For example, instead of using colors to indicate the escape speed of points outside the Mandelbrot, we could give them "height" — with higher points being closer to the Mandelbrot. Points belonging to the Mandelbrot set would have the highest height. Plotted this way, the Mandelbrot map looks like a flat plateau rising above the surrounding valley. This does produce a three-dimensional "terrain."

But this is no more intricate than the original 2D Mandelbrot; color has simply been replaced by a z-coordinate. The shape of the actual Mandelbrot map itself is still flat. Furthermore, to determine whether a point belongs to the Mandelbrot, we still need only two coordinates — x and y, where x is the real component of the number and y is the imaginary component. We want to know if there exists an algebraic map requiring a *three*-coordinate number to see whether a given number belongs.

The short answer is: no. The reason is simple: Numbers are inherently two-dimensional. All numbers that exist can be described by two values — one to indicate the real component and the other to indicate the imaginary component. For some numbers, one of those two values might be zero. But no third number is required. Therefore, all numbers can be plotted in the Argand plane; they require only two dimensions. Sets that are defined in terms of which numbers belong to them will inherently be two-dimensional.

We might ask, "Could numbers have a third, as-yet-undiscovered component?" Could there be a third axis to numbers that we have never used because in our experience it is always zero? After all, the realization that numbers have two dimensions rather than one is a relatively recent discovery. Throughout most of history, we only used numbers whose imaginary component was zero. In fact, over the course of history, human beings have consistently discovered new classes of numbers, unknown to their ancestors.

The History of Numbers

For example, at one time, people only considered the counting numbers. These are the positive integers (1, 2, 3, 4...). At some point, people thought it logical to include zero as a number. Not everyone agreed. It was counterintuitive to many that a number could represent nothing. For example, there is no Roman numeral for zero. Zero is now included in what we call the natural numbers.

But the natural numbers are not sufficient to describe all algebraic operations. Consider subtraction, for example. When we subtract 5 from 9, we get 4, which is a natural number. But what happens when we subtract 7 from 3? Although both numbers are natural numbers, the result is not. The subtraction operator suggested that numbers were not limited to the natural numbers. Hence, negative numbers were discovered. Again, it may seem counterintuitive at first to have a quantity that is less than nothing.

But without negative numbers, certain algebraic expressions would have no answer. We now readily accept that integers can be positive, negative, or zero.

The division operator suggested that integers were not the only type of number. While 8 divided by 4 yields the integer 2, division of other integers does not. The number 1 divided by 3 yields a number that is not an integer. Hence, rational numbers were discovered. These are necessary to describe fractions — quantities in between the integers. Rational numbers can be expressed as a/b, where a and b are each an integer. Without rational numbers, certain algebraic operations would have no answer.

The ancient Greeks were enamored with rational numbers. But they also liked geometry. The Pythagorean theorem required the Greeks to know about squaring a number. To find the hypotenuse of a right triangle from the other two legs, the Greeks had to square both legs and add the result. Then they had to find the square *root* of that result. The square root operator led to the discovery of a new class of numbers.

For example, the square root of 2 cannot be expressed as the ratio of two integers. It is not a rational number, and this can be mathematically proved. The Greeks — reluctantly — discovered this. This was the first time in history that a mathematical truth had been discovered that was completely unexpected and undesired. (This effectively disproves the hypothesis that human beings invented math). The Greeks were comfortable with rational numbers. But without the irrational numbers, many square roots of rational numbers would have no answer.

The combination of all rational and all irrational numbers constitutes the "real" numbers. The real numbers are inherently one-dimensional. Any position along a line can be expressed by a real number. The real numbers allow us to find solutions to almost any algebraic operator. Addition, subtraction, multiplication, and division of real numbers always results in a real number. But again, the square root operator implied that numbers exist besides the real numbers.

While the square root of a positive real number always yields a real number, the square root of a negative number does not. Since the square of any real number is never negative, the square root of a negative number cannot be a real number. Hence, the square root operator implies that a class of numbers exist that are neither positive, nor negative, nor zero. This led to the discovery of the so-called imaginary numbers.

Since imaginary numbers are not positive, they are not to the right of zero on the number line. Since they are not negative, they are not to the left of zero. And they are not zero. Thus, they are not on the real number line but are orthogonal to it, as we saw in chapter 1. The addition operator allows us to add a pure imaginary number to a real number, resulting in a number with a real part and an imaginary part, which we now call a complex number.

The details of the mathematics of imaginary and complex numbers were not discovered until the late 16th century. And even then, many people regarded the complex numbers as fictitious or useless. It wasn't until the late 18th and early 19th centuries that the power and usefulness of complex numbers was really appreciated.

There are two takeaways from this brief discussion of mathematical history. First, all known numbers were considered to be one-dimensional until the 16th century. All known numbers belonged to the real number line. But the complex numbers had not yet been discovered. Complex numbers

> Hence, the square root operator implies that a class of numbers exist which are neither positive, nor negative, nor zero. This led to the discovery of the so-called imaginary numbers.

are two-dimensional. Second, the existence of each new class of number was discovered algebraically. That is, algebraic operations on positive integers led to the discovery of negative integers, algebraic operations on integers led to the discovery of rational numbers, algebraic operations on rational numbers led to the discovery of irrational numbers, and algebraic operations on real numbers led to the discovery of complex numbers.

So, we must ask, do algebraic operations on complex numbers imply the existence of some greater class? The answer is no. The reason is simple: every algebraic operation on a complex number yields a complex number. The complex numbers are algebraically closed. So, if you pick any point in the Argand plane and perform any mathematical operation on it with any other point on the Argand plane, the result will be a point in the Argand plane. We now know of all the numbers required for any algebraic operation. Numbers have exactly two dimensions and no more.

To some, the fact that numbers are inherently two dimensional may be disappointing. After all, we live in a universe with three dimensions of space. And it might have been nice if numbers also were three-dimensional. But they aren't. There are certain physics problems, such as in fluid dynamics, that can be readily solved by applying complex numbers — but only in cases where only two dimensions are involved. Nonetheless, there are ways to apply mathematical rules to our three-dimensional universe.

> We now know of all the numbers required for any algebraic operation. Numbers have exactly two dimensions and no more.

Quaternions

One of the first methods that extended rules of mathematics to more than two dimensions was the method of quaternions. A quaternion is a conceptual entity very similar to a number but with two differences. First, a quaternion does not obey the commutative law of multiplication. Recall that for two numbers, a and b, the product of these two is independent of their order. That is, a × b = b × a. But for quaternions, this is not necessarily the case. Second, whereas numbers are inherently two-dimensional, quaternions are inherently four-dimensional. It is this second property that has made quaternions such a useful concept. In all other respects, quaternions behave just like numbers.

To be clear, quaternions are *not* numbers. But they behave just like numbers in all instances, except that they are not commutative under multiplication and are four-dimensional. While a number has a real part and an imaginary part (represented by a real number multiplied by i), a quaternion has a real part and *three* imaginary parts (represented by a real number multiplied by i, j, and k, respectively). As with numbers, the i represents the vertical y axis. But we also have j to represent a position in the z-axis and k to represent a position along a fourth axis that we cannot plot (or really imagine) in our three-dimensional space. Thus, any quaternion can be represented in the form a + b**i** + c**j** + d**k**, where a, b, c, and d are real numbers and **i**, **j**, and **k** are the orthogonal unit directions.

Additional rules were constructed to define how the extra two dimensions relate to each other. For example, i × j = k, and j × i = -k (remember, quaternions are noncommutative). But since they obey all other rules of algebra, quaternions can be used in place of numbers to solve physics problems in three-dimensional space. (Since quaternions are 4D, when only three dimensions are needed, one of the four components can be set to zero). Quaternions were introduced in 1843. They were found to be very useful in solving physics problems. However, by the end of the 19th century, most physicists began using vectors instead.

Vectors

A vector is simply a combination of numbers — usually real numbers. For a vector to be useful, we must also specify what each of the numbers means. As one example, a *position vector* describes a point in space using three numbers, the x, y, and z coordinates respectively. Since a vector can have any number of elements, it can describe a position in any number of dimensions. Often, vectors used in physics have three elements, one for each dimension of space.

But since they are not themselves numbers, most mathematical operations cannot be applied to vectors without modification. For example, what does it mean to multiply one vector by another? Do we multiply the first element in one vector by all the elements in the other, or just the first? Is the result also a vector or simply a single number?

Consequently, mathematicians have defined new operators to specify the relationships between vectors. For example, vectors can be multiplied using a *dot product* or a *cross product*, and the answer will depend on which operator is used.

Most physicists find vector notation much easier to use than quaternions. And the applications are endless.

> Most physicists find vector notation much easier to use than quaternions. And the applications are endless.

Quest for the 3D Mandelbrot

This background information brings us to a fascinating area of ongoing research: the search for the three-dimensional version of the Mandelbrot set. Just as there is a Sierpinski tetrahedron, which is the 3D equivalent of the 2D Sierpinski triangle, and a Menger sponge, which is the 3D equivalent of the 2D Sierpinski carpet, mathematicians have asked if there might be a 3D version of the Mandelbrot set.

We don't know in advance exactly what the 3D version *should* look like. But presumably, it should resemble the 2D version in its basic properties. Perhaps the cardioid will become an apple-shaped structure, and perhaps the circles growing off of it will become spheres. We expect many dendrites to extend away, but branching in more than just two dimensions. In any event, we expect that a slice down the middle of the x-y plane of this 3D shape should be the original 2D Mandelbrot.

Whatever this set is, it will not be defined as a set of *numbers*. This is because numbers are inherently two-dimensional, and hence their map must lie within the Argand plane. But we have now seen that we can use either quaternions or vectors to explore mathematical puzzles in three (or more) dimensions.

The natural choice to start with is quaternions because mathematical operators work the same way with quaternions as they do with numbers. Therefore, we can use exactly the same formula we used with the Mandelbrot set, and it will be perfectly well-defined for quaternions as well. Granted, quaternions are four-dimensional, and we require only a three-dimensional shape. But we can disregard the fourth dimension by investigating only the space in which the k-component is zero. So what happens when we run quaternions through the Mandelbrot iteration?

$$z_n^2 + c = z_{n+1}$$

The result is rather disappointing. We do get a volume of space representing those quaternions that are bound under the iteration. But rather than looking more complicated and interesting than the 2D Mandelbrot set, the result actually appears smooth — see figure 11.7. In fact, the resulting map is simply a "Mandelbrot of revolution." This is the result of rotating the 2D Mandelbrot map about the x-axis. The colors here represent the steepness of the change in direction at the boundary of the set — not escape values.

Although the complexity of the original is still mathematically present, it is not easily visible, having been smeared around the real number axis. No new complexity is gained. In fact, every plane that cuts through the x-axis contains exactly the 2D Mandelbrot and nothing more.

This new shape does satisfy our criterion that a slice down the x-y axis is the 2D Mandelbrot. And it is 3D. However, the circles have not become spheres. There are no "dendrites" because they have all been spread into cylindrical shapes. This 3D shape is simply a 2D Mandelbrot that has been rotated around an axis. Mathematically, this is because the j quaternion unit acts just like the i. That is, $j^2 = -1$. Therefore, there can be no genuinely new structures in this shape that are not in the 2D Mandelbrot. Most fractal enthusiasts agree that this shape should not be considered the true 3D Mandelbrot.

So, we turn to vectors. We can use vectors of three elements, one to represent each coordinate in three-dimensional space. But when we go to apply the Mandelbrot formula,

Figure 11.7

we immediately run into a problem. The Mandelbrot formula is for numbers and is not defined for vectors. For example, the formula involves taking a square. But how do we multiply a vector by itself? Recall we have both dot products and cross products for vectors. A dot product of a vector yields a single number, not a vector. And then it is not obvious how to add that number to "c," which is now a vector. On the other hand, the cross product of a vector with itself does yield a vector, but all its elements are zero. Neither definition is adequate.

A new definition of the power of a vector v was proposed by mathematician Paul Nylander. It is as follows:

$$v^n = r^n(\sin(n\theta)\cos(n\varphi), \sin(n\theta)\sin(n\varphi), \cos(n\theta))$$

where r is the total magnitude of the vector, and θ and φ are the angles between its components in spherical coordinates. We need not be concerned with the details of the formula. The point is that his formula made it possible to apply the Mandelbrot formula to vectors.

In 2007, Daniel White applied the vector power formula to the standard Mandelbrot formula. The result is shown in figure 11.8. In the x-y plane, the shape is the standard Mandelbrot set, which satisfies our expectation for the 3D version. And we now have new structure in the z-direction. So, this is not simply a rotated 2D Mandelbrot. It has structure in all three dimensions.

But is the structure fractal in the z-dimension? Note that this new shape does not match our expectation that circles in the Mandelbrot will become spheres or that the cardioid will become "apple" shaped. Large sections of this geometric solid appear quite smooth, stretched like taffy.

Figure 11.9 is a plot of the upper region of this new shape, approximately corresponding to figure 1.3 of the Mandelbrot set. We note that there is no clear evidence of scale-invariance in the z-axis. Since the Mandelbrot had circles upon circles upon circles, we were expecting to see spheres upon spheres upon spheres. But we don't. This new shape, while very interesting, is apparently not the 3D equivalent of the Mandelbrot set.

Figure 11.8

Figure 11.9

Nylander and White then began experimenting with higher powers of the Mandelbrot formula, as with the multibrots we explored in chapters 6 and 7. But if the vector version of the Mandelbrot doesn't produce a true 3D fractal, then surely the vector version of a multibrot won't either, right? One thing we've seen in exploring fractals is that mathematics does not always match our expectations. When we plot those position vectors that are bound under the formula for a multibrot of power 8 (e.g., z^8+c), the following shape results (figure 11.10).

Figure 11.10

Mandelbulb

Discovered in 2009, this new shape is called a Mandelbulb. It has some fascinating properties that mimic the Mandelbrot set or, more accurately, a multibrot of the eighth power. Notice that the overall shape is roughly spherical — a "bulb." And budding off of the surface of this bulb are multiple bulbs. And on each of these smaller bulbs, we see hints of even smaller bulbs.

Zooming in on the bulb just left of center, we obtain figure 11.11. We note that this bump has smaller bumps along its surface, which have smaller bumps along their surfaces. This shape has 3D structures that repeat infinitely on increasingly smaller scales. The Mandelbulb is a true 3D fractal! Powers higher than 8 also produce true 3D fractals.

But is the Mandelbulb the real 3D version of the *Mandelbrot set*? The overall shape is quite different. The Mandelbulb is round, whereas the Mandelbrot set is not. The Mandelbulb lacks the antenna of the Mandelbrot. And the formula is different; the Mandelbrot is a second-power formula, whereas the Mandelbulb is 8th power. The Mandelbulb does have spheres upon spheres, just as the Mandelbrot has circles upon circles. But a slice of the Mandelbulb in the x-y plane is not the Mandelbrot, but rather resembles a multibrot of power 8. At best, the Mandelbulb could be the 3D version of an 8th-power multibrot.

But there are other reasons to think that the Mandelbulb is not the 3D Mandelbrot. The Mandelbrot perimeter is infinitely wiggly and therefore has fractal structure at all locations. However, the Mandelbulb surface is fractal in only some locations. Other locations are smooth and have a taffy-like appearance. You can see some of these smooth regions in between the bulbs in figure 11.10.

The Mandelbulb seems to be the closest shape to a 3D Mandelbrot so far discovered. And it is a remarkable structure, worthy of exploration. But it's not quite a 3D version of the Mandelbrot. So the search for the true 3D Mandelbrot set continues. Given the rapid pace of discovery in this branch of mathematics, due partially to the increasing power of computers, the

discovery may be right around the corner. Perhaps a future edition of this book will include what all agree is the true 3D version of the Mandelbrot set. Then again, perhaps someone will demonstrate that such a shape is mathematically impossible. Mathematics is such an exciting field of discovery. And this is because we are exploring how God thinks about numbers. God's mind is truly amazing.

We have investigated fractals that are algebraically defined and fractals that are geometrically designed. However, there is yet another variety of fractal that we have yet to explore. The implications of this third class are devastating to any non-biblical worldview.

Figure 11.11

Physical Fractals and the Grand Dilemma

All the fractals we have investigated up to this point exist only in the conceptual world of mathematics. They are non-physical, not made of atoms. You cannot touch one. They do not have a particular location in the physical universe. As such, you cannot see one in a telescope. They exist in the mind. We can make physical plots of these fractals on a computer screen or on a piece of paper. But the shapes themselves exist in the mind. Ultimately, they exist in the mind of God. And through the gift of mathematics, God has revealed some of them to us.

However, there is also a class of fractals that *are* physical — fractals that are made of atoms. These physical fractals do have a location in space and can be touched. And they are found just about everywhere you look. In this chapter, we explore physical fractals and their relationship with mathematical/conceptual fractals. We will find that the relationship between these two fundamentally different types of fractals is a confirmation of the biblical worldview and fundamentally irreconcilable with any non-biblical position.

Our first example of a physical fractal is no doubt familiar to many: a snowflake (figure 12.1). Snowflakes are made of frozen water molecules, but they resemble the conceptual geometric shape we explored in the last chapter: the Koch snowflake (figure 11.1D). Like the Koch, the perimeter of a typical snowflake is wiggly, with structures that

Figure 12.1

repeat on increasingly smaller scales. Also like the Koch, the overall shape of a snowflake is generally six sided. The cause in this case is the shape of the water molecule.

There is an important difference, however. You can zoom in on the perimeter of a Koch snowflake forever, and the resulting plot will continue to exhibit complexity; the perimeter is infinitely wiggly.

However, you can only zoom in on a physical snowflake a finite number of times before the edge is no longer fractal. This is not merely a limitation of microscope technology. Rather, the fractal nature of a physical snowflake is finite and thus does not continue forever. In continuing to zoom in, at some point we would reach the level of the individual water molecules, which are non-fractal. The atoms that comprise the molecules can be approximated as spheres, with no fractal structure.

Figure 12.2

This is a common feature of *all* physical fractals. Unlike mathematical/conceptual fractals, which continue to exhibit scale-invariance on smaller scales without end, physical fractals do not. Physical fractals are scale-invariant for a *finite* number of scales and *cease to be fractal on smaller scales*. This is a necessary consequence of physical fractals being composed of atoms, which themselves are non-fractal.

Lightning is typically fractal in nature. The primary bolt often branches into secondary bolts, which then further branch and so on (see figure 12.2). Each smaller branch resembles the entire structure. Apart from context, we would have difficulty determining whether we are looking at the entire lightning bolt, a small portion, or an electric spark. This is another example of scale-invariance. The branching resembles the dendrites of the Mandelbrot set as seen in figure 1.4.

Another natural example of fractal branching occurs in trees. Many trees have a primary trunk that sprouts branches, each of which sprouts additional branches and so on up to some limit. This is often apparent in deciduous trees after they drop their leaves in the autumn (see figure 12.3).

Figure 12.3

In fact, many plants exhibit fractal properties. Consider the fern. It has a primary stem with leaves growing off the left and right sides (figure 12.4). But further inspection of each leaf shows that it too has a central stem, with leaflets growing off the left and right. Each leaf of a fern resembles the entire fern.

A particularly interesting example of fractals in the plant world is found in Romanesco broccoli (figure 12.5). The basic shape of the bulb is that of a cone. But this cone is made of smaller cones, which are made of smaller cones, which are made of smaller cones (figure 12.6). This is a 3D geometric fractal: one that is found in the physical world.

Fractals are even found in animals. Consider the chambered nautilus (figure 12.7). Each chamber is a larger or smaller version of the adjacent chamber. The overall structure is a single spiral. The circulatory system of living animals and human beings is fractal. Arteries branch into smaller branches, eventually branching into the smallest capillaries.

Figure 12.4

Figure 12.5

PHYSICAL FRACTALS AND THE GRAND DILEMMA

Figure 12.6

Geological features often exhibit fractal properties. The way rivers branch is fractal. This can happen upstream when we consider the way tributaries join to form larger streams. The fractal structure can also occur downstream in the delta formed as the river reaches the ocean. The delta formed by the Ganges, Brahmaputra, and Meghna Rivers in Bangladesh is a spectacular example of this (figure 12.8).

Figure 12.7

Figure 12.8

Figure 12.9

Figure 12.10

Mountains have a fractal structure. The largest mountains are made of smaller ones, which are made of smaller ones and so on. This is particularly evident in pictures of mountains taken from space. Figure 12.9 shows a section of the Himalayas taken from the International Space Station.

Clouds often exhibit fractal structure. Many types of clouds are made of smaller clouds, which resemble the basic structure of the larger version (see figure 12.10).

Physical fractals are not confined to our terrestrial sphere. The cosmos contains examples as well. Consider spiral galaxies. There are three basic types of spiral galaxies: flocculant, multi-arm, and grand design spirals. All three are fractal. Consider the Whirlpool Galaxy in figure 12.11. This is a grand design spiral because it has two primary spiral arms. It is therefore a double spiral very reminiscent of the structures in Double Spiral Valley of the Mandelbrot set. Compare figure 12.11 with figure 3.5. Galaxies themselves come in clusters, and there are hints that the way galaxies cluster may also be fractal.

Figure 12.11

The Grand Dilemma

In previous chapters, we found that fractals are built into math. When we map those numbers that belong to certain sets, the result is a structure of remarkable beauty and complexity. Yet, such maps are abstract. These fractals are not made of atoms and cannot be found anywhere in the physical universe. They exist in the mind.

However, we have seen in this chapter that fractals can also be found in the physical world. There are fractals that do have a location in space and can be touched. In many cases, they strongly resemble those fractals that are abstract. But how can this be? The similarities are far too great to be merely a coincidence. How is it that the physical universe bears such a strong resemblance to the abstract world of mathematical concepts?

We must admit that our thoughts do not control reality. You can try to make the sun set in the east, but it won't work. You can imagine that the sky is a different color, but it will have no effect on the actual color of the sky. Our thoughts have no power on the physical world. And yet, for some reason, the physical world has fractals that strongly resemble those fractals that exist only in the mind — those that are the result of a mathematical plot. How can we explain this?

One explanation is that the physical world obeys mathematical laws. The forces of gravity and electromagnetism that act on matter can be articulated by simple mathematical expressions. The simple formula $E = mc^2$ discovered by Einstein relates the energy of any object to its mass and the speed of light.

Likewise, electromagnetic forces that hold atoms together are expressible in mathematical terms. Snowflakes are generally six sided because of the shape of the water molecule as determined by the forces between the atoms. Since fractals are a natural result of the rules of mathematics, and since the physical universe obeys rules of mathematics, it stands to reason that the physical universe will have fractals. Mystery solved.

But is it? No doubt physical fractals occur because the physical world obeys mathematical laws and mathematical truths contain fractal beauty within them. But this explanation is only sensible if we can justify our expectation that the physical universe should obey mathematical laws.

In other words, of course fractals (which occur in the abstract world of math) can occur in physical reality since physical reality obeys math. But this merely pushes back the question. We must now ask: why? *Why* does the physical world obey mathematical laws?

That the physical universe does obey mathematical laws is not in question. We all assume that it does. The question is: why? How does *physical reality* (which has no apparent connection to mental activity) know anything about mental concepts like numbers? What worldview can justify our expectation that physical reality obeys conceptual laws?

Many people are inclined to simply dismiss questions that are difficult. "Who cares why the universe obeys mathematics? It is sufficient that it does." But deep thinkers are not satisfied to see a connection and simply dismiss it as irrelevant. Obviously, there is a reason *why physical reality obeys conceptual truths*, and the answer may have profound implications. I suggest that no non-biblical worldview can account for the connection between mathematical truths and physical reality.

We may not be bothered by the dilemma because we simply take for granted that the physical universe obeys mathematical laws. We are used to it. We tend to not question those things with which we have familiarity. But our familiarity with a topic is not the same as an explanation for it.

After all, we are familiar with gravity, but who can really explain it? How does a rock know which way to fall? How does it know where the earth is so that it can fall toward it? It has no eyes to see where earth is, yet

> How does *physical* reality (which has no apparent connection to mental activity) know anything about mental concepts like numbers? What worldview can justify our expectation that physical reality obeys conceptual laws?

it always falls toward it. We say there is an invisible force that reaches from the earth and pulls on the rock. But is this really an explanation or merely an observation?

Likewise, we accept that physical reality obeys mathematical truths. But few have stopped to consider how strange that situation really is. Mathematics occurs in the mind. It is entirely abstract. How does the physical universe know what goes on in our minds such that it can obey?

Perhaps someone will suggest that mathematics is derived from the physical universe — that our discoveries of mathematics are based on physical reality. Such a person will say, "Of course the physical universe obeys mathematical laws because we discovered mathematics by looking at the physical universe. Math is simply the name we give to the way the universe behaves." But this answer is naïve and not true to the history of mathematics. As one example, the complex numbers were not discovered by observing the physical universe. They were discovered by pure reasoning.

Only later did we discover that they have application to the physical world.

Many mathematical truths have no parallel in the physical world — a fact that would be incomprehensible if mathematics merely described physical reality. For example, in mathematics we can compute the higher-dimensional equivalent of volume for geometric shapes in dimensions higher than three. As far as we know, our physical universe has only three dimensions of space (and one of time). Yet, this doesn't prevent us from discovering mathematical truths in higher numbers of dimensions.

In fact, some theories in the field of quantum mechanics postulate that there may be additional dimensions of space that are normally "hidden" from view. No one doubts that if such extra dimensions are eventually discovered, they will obey laws of mathematics. But why should we assume that? Clearly, such higher-dimensional math was not discovered as a result of observing the physical universe. Rather, the math was discovered first. And if the physical universe has such extra dimensions, we expect them to obey the conceptual mathematical truths we have already discovered regarding higher dimensions. But why expect this?

Indeed, mathematics is the logic of numbers. Both numbers and logic are abstract and have no physical substance. If our minds could exist with no physical universe at all, it is entirely conceivable that we could have still discovered mathematical truths, for such truths do not depend on the physical world. Yet, for some reason, the physical world obeys mathematical truths. But why?

This issue is not a mere triviality. Rather it has plagued secular thinkers for quite some time. In 1960, the Nobel-prize-winning physicist Eugene Wigner (1902–1995) tackled this dilemma in a paper entitled "The Unreasonable Effectiveness of Mathematics in the Natural Sciences."[1] From his secular perspective, Wigner points out that "the enormous usefulness of mathematics in the natural sciences is something bordering on the mysterious and that there is no rational explanation for it."

Indeed, Wigner's secular worldview cannot even account for the human mind's capacity to reason mathematically. I appreciate his honesty when he states,

> Why should the physical universe obey laws at all? In a chance universe, one that is not controlled by any rational mind, why should we expect to find rational patterns?

The great mathematician fully, almost ruthlessly, exploits the domain of permissible reasoning and skirts the impermissible. That his recklessness does not lead him into a morass of contradictions is a miracle in itself: certainly it is hard to believe that our reasoning power was brought, by Darwin's process of natural selection, to the perfection which it seems to possess. However, this is not our present subject.

Why should the physical universe obey laws at all? In a chance universe, one that is not controlled by any rational mind, why

> "...the enormous usefulness of mathematics in the natural sciences is something bordering on the mysterious and that there is no rational explanation for it."
> – Eugene Wigner

---

[1] Eugene Wigner, "The Unreasonable Effectiveness of Mathematics in the Natural Sciences," in *Communications in Pure and Applied Mathematics*, vol. 13, No. I (February 1960) New York: John Wiley & Sons, Inc. Copyright © 1960 by John Wiley & Sons, Inc.

> "The miracle of the appropriateness of the language of mathematics for the formulation of the laws of physics is a wonderful gift which we neither understand nor deserve."
>
> – Eugene Wigner

should we expect to find rational patterns? In a universe in which things are constantly changing, why should we find invariant truths? This in itself is bothersome. But it is even more remarkable that the laws the universe does obey should be mathematical in nature and congenial to human understanding.

Wigner expresses these two remarkable surprises as follows:

> It is difficult to avoid the impression that a miracle confronts us here, quite comparable in its striking nature to the miracle that the human mind can string a thousand arguments together without getting itself into contradictions, or to the two miracles of the existence of laws of nature and of the human mind's capacity to divine them.

Wigner's analysis is very insightful. Few people have so eloquently expressed the grand dilemma. Indeed, few people have given the issue much thought. We trod along, confident in our expectation that laws of nature will continue to be discovered by application of mathematics, completely unconcerned as to the miracle that makes such discoveries possible. We are like a bird who eats from a birdfeeder without giving any thought to how the food got there.

What then is the conclusion of Wigner's analysis? He closes with the following paragraph:

> Let me end on a more cheerful note. The miracle of the appropriateness of the language of mathematics for the formulation of the laws of physics is a wonderful gift which we neither understand nor deserve. We should be grateful for it and hope that it will remain valid in future research and that it will extend, for better or for worse, to our pleasure, even though perhaps also to our bafflement, to wide branches of learning.

Wigner is grateful for the appropriateness of mathematics to natural science. But to *whom* is he grateful? Being grateful implies that someone has shown grace — extended unmerited favor. Wigner hopes that mathematics will continue to prompt new discoveries in physics, but he has *no rational basis* for such hope on his professed, secular worldview. Indeed, he admits that he cannot even account for the *current* success of science, to say nothing of his confidence in any future success.

The problem is not with Wigner's intelligence or insight. He was a truly brilliant man, and his analysis on this topic is ingenious — insofar as is possible from a secular perspective. The problem is his worldview. A secular worldview cannot make sense of the relationship between mathematics and science because the secular worldview is fundamentally wrong.

Mathematics is the logic of numbers, the study of the rules regarding the relationship between numbers. As such, mathematics is an abstract endeavor of the mind. Without a mind, there can be no mathematics. And since the universe is governed by mathematical laws, it is necessarily the case that the universe is controlled by a rational mind.

This, of course, is the Christian position. We have already seen in chapter 8 that the Christian worldview makes sense of mathematics and mathematical properties. Mathematical truths are universal, invariant, abstract entities without exception, precisely because God is an omni-present, unchanging, sovereign Being who thinks. And since God has revealed some of His thoughts to us, we are able to discover some mathematical truths.

God is also sovereign over the physical universe. He upholds all things by the expression of His power (Hebrews 1:3). From creation, He has determined what will happen in His universe (Isaiah 46:9–10). Since God's

> Without a mind, there can be no mathematics. And since the universe is governed by mathematical laws, it is necessarily the case that the universe is controlled by a rational mind.

mind controls the physical universe, and since God thinks mathematically, it follows that the universe will obey mathematical laws. This grand dilemma for the secularist is a necessary and logical truth for the Christian.

Furthermore, God created human beings in His own image, after His likeness (Genesis 1:26–27). As such, human beings have a finite capacity to think in a way that is consistent with the character of God. And God commands us to do so (Isaiah 1:18, 55:7–8; Psalm 32:9). This includes (but is not limited to) our ability to think mathematically. So it stands to reason that we should be able to understand, at least in part, the systematic way that God upholds creation by studying mathematics. The biblical worldview makes sense of the effectiveness of mathematics in the nature sciences.

Hence, our answer to the grand dilemma is conceptually simple. Fractals exist both in the abstract world of mathematical thought and also in physical reality because God's mind controls both. The thoughts of the Lord are beautiful, and we find such beauty when analyzing His thinking about numbers and also in the physical world that He created.

Critics who fail to grasp the basics of such epistemological truths will sometimes claim that this answer is insufficient, being simply a "god-of-the-gaps" approach. Namely, some people will say that our response to any mystery is simply to say, "God did it." And furthermore, they claim that the phrase "God did it" can answer any question but doesn't really provide any helpful information.

But this isn't the case, at least not with the God described in the Bible. The biblical God has certain very specific characteristics that He has revealed to us in His Word. For example, God is self-consistent and cannot deny Himself (2 Timothy 2:13). Hence, if we discovered a contradiction in nature, two truths that are contradictory to each other, the biblical worldview *could not account for this*. We could not truthfully say "God did it," because the Bible says that God cannot do that!

Furthermore, if the universe was truly haphazard with no patterns at all, the Christian worldview would not be able to make sense of this. Indeed, it would then be the atheistic worldview that would better fit such a hypothetical universe. If all human beings were utterly incapable of discovering any truth, then this would be impossible to reconcile with the Christian worldview in which God has revealed some truth to us. Saying "God did it" would not resolve such a paradox because the biblical God is not that way.

So, we are not inserting God to resolve a dilemma. Rather, we recognize that there is no dilemma, given the nature of God, man, and the universe as described in the Bible. Given what the Bible teaches, the universe *must* be mathematical in its behavior. Given the nature of man according to Scripture, he will have a limited capacity to discover mathematical truths, both abstract and with physical application.

It is not what we *fail* to understand about the universe that demonstrates the existence of the biblical God, but rather what we *do* understand. The physical universe does obey mathematical laws, and this is expected by the Christian, but it challenges the secularist.

Furthermore, non-biblical religions cannot make sense of the appropriateness of the language of mathematics for the formulation of physical laws. We have already seen in chapter 8 that polytheistic religions cannot make sense of mathematics at all. Furthermore, any god besides the biblical one would not have the properties necessary to justify the properties of mathematical truths (universal, invariant, and without exception) and their application to the physical world. It is Jesus, and no other god, who upholds all things by the expression of His power (Hebrew 1:3). The biblical God is an omni-present, unchanging, sovereign Spirit whose mind determines truth. And we know these things because God has revealed them to us objectively in the Bible.

> It is not what we *fail* to understand about the universe that demonstrates the existence of the biblical God, but rather what we *do* understand.

About the best any non-Christian could do would be to attempt to create a god who is a copy of the Christian God. "I can also make sense of mathematics and its application to the physical world. My god is just like yours: omni-present, unchanging, all powerful, and so on. So I don't need to be a Christian to make rational sense of mathematics." But this answer fails for at least two reasons.

First, any god who is exactly like the Christian God would be the Christian God. It is the characteristics of the biblical God that justify mathematics and its physical application. If your God does this, then He is the God of the Bible, and you need to bow to Him, repent of your sins, and acknowledge Him as Lord.

Of course, the person may say that their god is almost like the biblical God, except that he did not write the Bible. And this brings us to our second refutation. Without the Bible, how can you know anything about your god? A person might claim that their god privately revealed himself/herself/itself, giving all sorts of revelation. But then I'm going to ask, "How do you know it was truly revelation from a god and not your own imagination?"

The Living God gave us objective revelation in the Bible. God used over 40 human authors to pen His Word, thus ensuring that it is not simply a human opinion. It is objective because it is open to inspection by anyone. A private "revelation" is not. If God had revealed Himself only through private revelation, then we could never be sure that we knew anything about God. So any made-up god can never be known because there is no objective revelation. The biblical God alone can make sense of the complexity and beauty we find in fractals.

Hence, fractals, along with everything else, are a demonstration of the truth of the biblical worldview. The images in this book are not the result of human creativity. The only human element is the selected color scheme, but the shapes stem from the mind of God. What a mind!

God has revealed Himself to us in many ways. He has written His moral standard into the core of our being (Romans 2:14–15) by which we have knowledge of right and wrong. He has hardwired knowledge of Himself into us such that when we look at the creation, we instantly recognize it as the Word of God (Romans 1:18–20). But since we have a tendency to distort such internal knowledge, God has given us objective, written revelation — the Bible.

The Bible tells us much about God's nature. He is all-powerful, having spoken the entire universe into existence. He is sovereign, controlling every atom in every molecule of the cosmos. He is all-knowing; we have seen a glimpse of the infinite nature of His mind as revealed in fractals. And God is just; He judges rightly.

When we take all these facts together, we realize that we have a horrible problem. We have not fully obeyed the commandments of the King of kings. We have committed treason against Him. There is no doubt that God is aware of this, since He is all-knowing. And since God is a righteous judge, He will not let crime go unpunished. Treason is a capital punishment, and we have rebelled against an infinite King. We therefore deserve an infinite death. And since God is all-powerful, there is no doubt that He will be able to carry out just punishment.

It is not surprising that people really do not want to believe in the biblical God.

It would be frightening enough to offend your earthly ruler. But in that case, you might be able to escape. And even if caught, you might hope to be found innocent in court. But to be in a position where an all-powerful, all-knowing Being is rightly angry at you, knowing that His justice ensures that you will be found guilty in court, knowing that your eternal punishment is assured and what you rightly deserve, is terrifying. People don't want to believe this horrifying truth. But it is true nonetheless.

But there is one other crucial aspect of God revealed in Scripture. He is *merciful*. For anyone who is genuinely regretful of their rebellion against Him, anyone who asks forgiveness and trusts in Him, God will save. He will forgive their rebellion. And so that justice may be satisfied, He Himself paid the penalty for our treason. It is impossible to fully grasp the depth of God's love — that He would become man for the purpose of dying a horrible death in our place so that we might be made righteous.

> It is impossible to fully grasp the depth of God's love — that He would become man for the purpose of dying a horrible death in our place, so that we might be made righteous.

This is the gospel. Jesus is the Son of God. As a man, He represented us on the Cross. As God, He could pay an infinite penalty. The same God who instilled such infinite beauty into mathematics was willing to die to pay our price. He rose again on the third day, proving that He has power over life and death. And He has offered salvation and forgiveness for all who repent and trust in Him, confessing Him as Lord.